Mycotoxin Contamination and Control

Edited by:

Henry Njapau
Food and Drug Administration
Center for Food Safety and
Applied Nutrition
College Park, Maryland, USA

Socrates Trujillo
Food and Drug Administration
Center for Food Safety and
Applied Nutrition
College Park, Maryland, USA

and

Albert E. Pohland
AOAC INTERNATIONAL
Governmental Affairs and
International Programs
Gaithersburg, Maryland, USA

Douglas L. Park
Food and Drug Administration
Center for Food Safety and
Applied Nutrition
College Park, Maryland, USA

authorHOUSE®

AuthorHouse™
1663 Liberty Drive, Suite 200
Bloomington, IN 47403
www.authorhouse.com
Phone: 1-800-839-8640

First published by AuthorHouse 2/28/2008

ISBN: 978-1-4343-3544-9 (sc)
ISBN: 978-1-4343-3545-6 (hc)

Library of Congress Control Number: 2008900364

Printed in the United States of America
Bloomington, Indiana

This book is printed on acid-free paper.

Proceedings of the International Workshop on Mycotoxins, held July 22-26, 2002 at the U.S. Food and Drug Administration, College Park, Maryland, USA.

INTERNATIONAL WORKSHOP ON MYCOTOXINS

PAHO * FAO * FDA * USDA * JIFSAN * IAEA * EC * WHO * UNEP * CSL-UK

Forward

The International Workshop on Mycotoxins, held July 22-26, 2002, at the U.S. Food and Drug Administration facility in College Park, Maryland, was neither the first of its kind nor will it be the last. It was, however, unique in its conceptual framework and theme: - "an attempt to harmonize mycotoxin programs worldwide." Although the development and blooming of this all-encompassing concept remains to be seen, the baby has taken the first steps and the required nourishment is being mobilized. The proceedings of the symposium are contained in this book

Unavoidable, naturally occurring toxins such as the mycotoxins pose a unique challenge to the worldwide goal of providing a safe, wholesome food supply. Despite improvements in agricultural production practices, mycotoxins continue to have a negative impact on the availability of safe food. According to the Food and Agriculture Organization (FAO), a substantial quantity of the world's food is contaminated with mycotoxins every year. Consequently, the significance of mycotoxins to food security and safety, and the increasing global trans-shipment of food commodities heighten the need for a concerted, worldwide effort to reduce mycotoxin contamination of foodstuffs.

As part of a broad international food safety outreach effort to assist economically challenged nations improve food safety and quality, as well as compliance with international trade standards, the U.S. Food and Drug Administration (FDA) and the Joint Institute for Food Safety and Applied Nutrition (JIFSAN), University of Maryland, initiated a series of activities to assist resource-poor countries minimize mycotoxin contamination of their foodstuffs. Initial activities involved two years of discussions with experts from various organizations including the European Commission (EC), World Health Organization (WHO), International Atomic Energy Agency (IAEA), the Food and Agriculture Organization (FAO), Pan American Health Organization (PAHO), United Nations Environment Program (UNEP), AOAC INTERNATIONAL, the U.S. Department of Agriculture (USDA), and Central Science Laboratories in the United Kingdom (CSL-UK). The outcome of the discussions was a five-year program to educate technocrats, policy makers and the general public about mycotoxins. The initial step was the conduct of a train the trainer workshop. The workshop consisted of plenary lectures by internationally recognized experts followed by "hands-on" laboratory exercises.

Recognizing that the development of effective and sustainable mycotoxin surveillance and prevention programs requires that all those involved are aware of the hazards associated with mycotoxins and appreciate their respective roles in minimizing them, pertinent elements of the training included a review of the conditions that lead to the formation of mycotoxins in food commodities and health risks associated with consuming mycotoxin contaminated foods. Other aspects of the training were: sampling, detection methods, risk assessment, establishment of regulations and monitoring programs and, compliance with international trade standards. Training materials used at the initial workshop were developed in a manner that allowed for their subsequent use at satellite regional training locations worldwide. Participants were encouraged to incorporate other training materials prepared by national and international organizations when conducting satellite workshops.

One hundred and twenty (120) individuals from 48 countries attended the plenary sessions; 54 participated in the hands-on laboratory exercises. The majority of the participants were aptly from African (18) and Asian (11) countries. The rest were from countries in Europe (9), South America (7) and North America (3). Participating individuals included government health officials, agricultural specialists, researchers, academicians and food technologists. Selection of participants was based on the severity or potential of mycotoxin contamination in the home country, technical qualifications, and the potential to accomplish the objectives of this effort upon their return.

Individuals that received the training returned to their respective regions/countries and were expected to embark on preparing to conduct similar workshops. Each participant would be provided with the plenary session monograph (in English, Spanish or French), a basic mycotoxins workbook/laboratory manual, FAO HACCP manuals, audio/visual VHS tapes (NTSC, PAL or SECAM), and compact discs containing all the lecture presentations. Where necessity could be demonstrated, 2" x 2" plastic slides of the lectures will be made available. In the last phase of this effort, the effectiveness of the outreach training program will be evaluated through a follow-up symposium at which participants will present the accomplishments of their respective training and mycotoxin control programs.

Mycotoxin contamination of food commodities can grossly affect a food exporting country's participation in global trade, because major food commodities such as corn, wheat, and peanuts and, to a lesser extent rice, are susceptible to mycotoxin contamination and may be subject to importing country's mycotoxin regulations. The regulations differ from one country to another and are non-existent in the majority of countries most affected by mycotoxin contamination. By promoting the coordination and adoption of similar mycotoxin detection and surveillance methods worldwide, this program is poised to significantly contribute to the effort to harmonize international food safety standards in a science-based and transparent manner in conformity with the Codex Alimentarius Commission's objectives of providing safe and wholesome food. Furthermore, improving the quality of food in economically challenged nations enhances the health and productivity of the populace, ultimately increasing economic growth and development. Richer economies in the developing regions will have increased purchasing and choice capabilities to participate in world trade.

Finally, it is our hope that the tunnel we travel is wide and the light at its end is as bright as the summer sun.

Editors

Acknowledgement

The workshop organizers are grateful to Mr. Joseph A. Levitt, Director, FDA Center for Food Safety and Applied Nutrition; Dr. David R. Lineback, Director, the Joint Institute for Food Safety and Applied Nutrition, University of Maryland; Dr. Terry C. Troxell, Director, FDA Office of Plant and Dairy Foods and Beverages, and Mr. Roy Barrett, USDA Foreign Agricultural Service for their encouragement and support. Further gratitude is extended to Mr. Barry Smith, Mr. Ray Banks, Ms. Christine Hileman and Mrs. Carole J. Shore (FDA), and Mrs. Mary Grimley (JIFSAN, University of Maryland/College Park). Appreciation is extended to the session chairs: Dr. Ramesh Bhat, National Institute of Nutrition, Indian Council of Medical Research, India; Dr. Dora N. Akunyili, National Agency for Food and Drug Administration and Control, Nigeria; Dr. Magda Carvajal, National Autonomous University of Mexico (UNAM), Mexico; and Dr. John Gilbert, Central Science Laboratory, U.K. for an excellent job. We would also like to thank the speakers and contributing authors, and the participants without whom the workshop and this book would not have been realized.

The FDA Office of Plant and Dairy Foods and Beverages and the Joint Institute for Food Safety and Applied Nutrition provided major funding for the workshop. Additional financial assistance was made available by USDA Foreign Agricultural Service (USDA/FAS), FAO, IAEA, WHO, EC, AOAC INTERNATIONAL, International Tree Nut Council/ Almond Board of California, CSL-UK, and PAHO.

The following industries, organizations and institutions provided direct and/or indirect financial and material support.

- Romer Labs Inc., (USA)

- Vicam Corp., (USA)

- R-Biopharm (USA)

- Agilent Technologies (USA)

- Peanut Collaborative Research Support Program (Peanut CRSP),

- United States Agency for International Development (USAID),

- International Organization for Chemical Sciences in Development (IOCD).

Organizing Committee

Douglas L. Park, Chair, FDA, USA
Elke Anklam, DG JRC, IHCP, Italy
Roy Barrett, USDA/FAS, USA
Ezzeddine Boutrif, FAO, Italy
Bruno Doko, IAEA, Austria
Jean-Marc Fremy, AFSSA, France
Hiremagalur N.B. Gopalan, UNEP, Kenya
David R. Lineback, JIFSAN, UMD, USA
Paisan Loaharanu, IAEA, Austria
Gerry Moy, WHO/IPCS, Switzerland
Samuel W. Page, WHO, Switzerland
Maya Pineiro, FAO, Italy
Albert E. Pohland, AOAC, USA
Terry C. Troxell, FDA, USA

Scientific Committee

Douglas L. Park, Chair, FDA
George A. Bean, UMD, USA
Elizabeth M. Calvey, FDA/JIFSAN, USA
Hans P. van Egmond, RIVM, The Netherlands
Robert M. Eppley, FDA, USA
John Gilbert, CSL, UK

Arthur J. Miller, FDA/JIFSAN, USA
Stanley Nesheim, FDA, USA
Henry Njapau, FDA, USA
Samuel W. Page, WHO, USA
Albert E. Pohland, AOAC, USA
Carole J. Shore, FDA, USA
Michael E. Stack, FDA, USA
Frederick S. Thomas, FDA, USA
Mary W. Trucksess, FDA, USA
Socrates Trujillo, FDA/JIFSAN, USA

Administrative and Technical Committee

Theodore L. Chambers, FDA
Mary A. Grimley, JIFSAN, UMD, USA
Christine L. Hileman, FDA/JIFSAN, USA
George J. Jackson, FDA, USA
John M. Lyczak, FDA, USA
Judy K. Quigley, JIFSAN, UMD, USA
Marcia. L. Meltzer, FDA, USA
Katherine Young, FDA, USA

Table of Contents

Section I: Overview

1. THE CONCEPT OF FOOD SAFETY

Douglas L. Park and Henry Njapau

Center for Food Safety and Applied Nutrition
Food and Drug Administration
5100 Paint Branch Parkway
College Park, Maryland 20740

Life, except for viruses perhaps, is sustained by feeding, the act of consuming food. According to *The Random House Dictionary of the English Language* (1), food is "any nourishing substance that is eaten, drunk, or otherwise taken into the body to sustain life, provide energy, promote growth, etc." Thus, water, if considered nourishing, can be classified as a food substance. Indeed, U.S. law makers may have included water in the Federal Food, Drug and Cosmetic Act, 201 (f), definition of food as "articles used for *food* for man or animals, ..., articles used for components of any such *article*" (emphasis added). According to this Act, water could be classified as a food article even if it is not nourishing, because it is used in the preparation of food.

Can the article that provides nourishment, called food, be unsafe? Ordinarily no, because experience has taught us to consume only those plant and animal products that have no obvious adverse effects, particularly if they manifested immediately. Hence, all living organisms instinctively ingest only those materials that cause minimal or no harm. However, food articles may be rendered unsafe by extrinsic substances such as naturally occurring fungal and bacterial metabolites, agrochemical contaminants, and in some instances food additives (Table 1.1). Ingestion of the actual pathogenic organism (bacteria, viruses, and parasites) may constitute a food-borne hazard as well (Table 1.2).

1

Table 1.1. Examples of naturally occuring and extrinsic chemical compounds of food safety concern

Class	Group	Name of compound	Commonly contaminated food
Naturally occurring	Plant constituents	Alkaloids, a-amanitin	Mushrooms, botanicals
	Mycotoxins	Aflatoxins, fumonisins, deoxynivalenol (DON)	Grains and grain products, oilseeds and nuts
	Aquatic biotoxins, Phycotoxins	Ciquatoxin, tetradotoxin, paralytic shellfish poisoning	Fish, shellfish, other seafoods
	Decomposition components	Scombroid poisoning	Tuna fish
Agrochemicals	Pesticides	DDT, rodenticides	Grains, fruits, vegetables, animal products
	Plant nutrients	Fertilizers	
	Animal drugs	Antibiotics, hormones	
Food additives	Direct	Preservatives, colors	Canned foods, processed products
	Indirect	Pesticide and drug residues	Grains, fruits, animal products

Table 1.2. Selected pathogenic organisms that affect food safety

Class	Pathogenic species	Affected food types
Bacteria	*Salmonella, Listeria monocytogenes, Vibrio cholerae, Clostridium botulinum, Escherichia coli, Shigella, Campylobacter jejuni, Yersinia, Staphylococcus aureus, Bacillus cereus, Vibrio vulnificus, Vibrio parahaemonlyticus*	Eggs, meat and meat products, water canned foods, fruits, vegetables, shellfish
Parasites	*Giardia lamblia, Cryptosporidium parvum, Cyclospora cayetanensis, Entameoba histolytica*	Water, fruits, vegetables
Viruses	Norwak, rotaviruses, hepatitis A	Shellfish, raw fruits and vegetables

In most countries, specific departments or agencies are charged with the responsibility of ensuring that substances whose purpose is nourishment provide only this desired effect. They ensure that food is safe. Because contaminated food may be hazardous, food safety programs comprise a combination of activities aimed at reducing or eliminating the associated risks. Therefore, food safety is the concept that, if implemented properly, results in a safe and wholesome food supply that meets nutritional needs. It is the driving force for the establishment of the activities that comprise risk assessment. The number of factors listed in Tables 1.1 and 1.2 shows how daunting the task of ensuring a hazard-free food supply can be. This chapter discusses the application of the concept of food safety and provides an overview of programs aimed at reducing the risks associated with mycotoxin contamination of human foods and animal feeds.

RISKS POSED BY MYCOTOXINS

Mycotoxins can contaminate human foods and animal feeds as a result of the growth of fungi (molds) before harvesting or because of improper storage. Preharvest contamination usually occurs when environmental conditions favor mold growth and mycotoxin formation in the field, whereas postharvest contamination is a function of poor storage. Preharvest conditions are, for the most part, beyond the control of humans. Aflatoxin B_1 has been recognized as a significant food-borne hazard since the Turkey X episode on English farms in 1960. The literature is replete with reports of serious animal (2, 3) and human (4–6) aflatoxicosis, sometimes resulting in death. Detailed discussions on conditions that favor fungal growth and mycotoxin production and the negative effects of mycotoxins on the health of humans and animals are presented in subsequent chapters.

Risk assessment constitutes a series of scientific activities executed in a coordinated manner to determine the significance of the risk posed by mycotoxins. The activities are categorized as: exposure assessment, toxicological determination, and hazard characterization (Table 1.3). The exposure phase determines levels of contamination in susceptible food commodities and consumption by target populations. Determining the toxicological hazard for the compound of interest involves identifying

3

the toxic event(s) and target organ(s), often in animals, and the relevance of the data to humans. The susceptibilities of subpopulations, such as the young, elderly, immune-suppressed, and nutritionally compromised may also be assessed at this stage. With this information, a numerical estimate of exposure (e.g. 2 µg aflatoxin/kg bw/person/year) can be determined and the probability of an adverse effect (disease event) occurring is estimated—i.e., hazard characterization.

Table 1.3. Activities that comprise a risk assessment exercise for a known toxic substance

Component	Activity
	-Identify susceptible commodity and the toxin
	-Select analytical procedure for toxin(s) of interest
	-Determine contamination levels, frequency, and seasonal/ annual fluctuations
Exposure assessment	-Identify target population and vulnerable subgroups within population
	-Determine intake of target commodity per capita
	-In vitro toxicity assessment of toxin(s)
	-In vivo toxicity assessment of toxin(s)
	-Modes of intake, distribution, and target organs
Toxicological determination	-Epidemiological evaluations
	-Animal to human extrapolation
	-Combine information from exposure assessment and toxicological determination to determine risk in terms of:
Risk characterization	-Dose-response assessment
	-No observed adverse effect levels

FOOD SAFETY MANAGEMENT PROGRAMS

The need to limit toxic substances in food and feed is based on two major concerns: (i) the adverse effects of the toxic substances on human and animal health, and (ii) the presence of the toxic substances or their metabolites in human foods derived from animals fed contaminated feed. There is ample evidence that, in the absence of efforts to curtail the occurrence of mycotoxins, food supplies can be substantially compromised (7). Because mycotoxins are naturally occurring toxic secondary mold metabolites, and their occurrence in foods is unavoidable and unpredictable, food safety control efforts focus on minimizing their presence to the greatest extent feasible.

Currently available procedures for reducing the presence of naturally occurring toxins in foods fall into two major classes: prevention of contamination and removal of the contaminant. Preventive measures are the primary means of managing hazards associated with mycotoxin contamination, and their effectiveness has been proven (8). In recent years, genetically modified plants and biotic control efforts have shown promising results (9, 10). In general, implementation of good agricultural practices and quality assurance programs, perhaps tailored to the principles of hazards analysis critical control point (HACCP) (11), where applicable, can ensure that agricultural commodities are minimally contaminated.

Additional aspects of management can be infused into the programs should mycotoxin contamination occur as a result of ineffective preventive measures. These efforts include establishing regulatory limits and monitoring as well as removing or minimizing mycotoxins through processing and/or decontamination procedures. Factors such as the public health significance of the hazard and the effect regulation would have on the economy and availability of food are considered when mycotoxin regulations are established. In some instances, legal infrastructure and government financial support, analytical capabilities, and consumer awareness play a significant role in sustaining such activities. These schemes are not restricted to the mycotoxins; they apply to all other substances that may contaminate foodstuffs.

REGULATIONS

Once the risk posed by a particular toxic substance has been established and determined to be of concern, exposure of the population to the toxic substance may be minimized by setting limits. The limit defines a contamination level above which consumption is forbidden or is highly discouraged. For instance, in the United States, the Commissioner of Food and Drug Administration (FDA) concluded that the observation of severe carcinogenic effects in experimental animals and reports of positive correlations between dietary aflatoxins and primary liver cancer in humans in other parts of the world were sufficient justification to take action to control human exposure to aflatoxin to the lowest level possible. 20 µg/kg was set as the maximum permissible level in human food commodities (with the exception of milk, 0.5 µg/kg) suspectible to aflatoxin contamination. Mycotoxin concentration is often expressed as micrograms per kilogram (µg/kg) or parts per billion (ppb): 1 ppb = 1 × 10^{-6} g of a mycotoxin per kilogram of product (Whitaker, T.B., personal communication). The FDA's authority to use the 20 µg/kg limit for enforcement action is based on section 402(a)(1) of the Food, Drug, and Cosmetic Act. Details of historical and scientific fact dictating the course of action taken by the FDA are discussed by Park and Stoloff (12) and summarized in chapter 22.

Numerous factors are considered when regulatory limits are set for naturally occurring toxins in foods and animal feeds. In the case of milk, the age of the consumer and the large quantities consumed per unit body weight play an important role in determining the limit. Similarly, the relatively high aflatoxin in feed to aflatoxin in milk ratio influences the setting of the action level for dairy feeds. Regulatory limits set without due scientific process may be meaningless and impossible to enforce. For example, the establishment of tolerance levels at 5 µg/kg or less by a number of countries (13) may be desirable from a hazard point of view. However, such a tolerance level may not be practical because mycotoxins are natural contaminants that often cannot be completely excluded from the food, and current analytical methods have substantial uncertainty at such low levels. Factors that affect the setting of regulatory limits include:

- Control of human exposure to mycotoxins

- The source of the mycotoxin contamination

- Toxicological characteristics of mycotoxin residues and their metabolites

- The capability of current methods to measure and confirm the identity of such residues

- The relationship between mycotoxin levels in feeds and their residues in animal products

- The effect of particular control measures on the availability of the food or feed

- Practicability and effectiveness of regulatory enforcement strategies

MONITORING

Monitoring is a surveillance activity initiated in response to an established hazard posed by an identified toxin(s) in a specified product(s) and/or commodity. It is also governed by prior knowledge about the susceptibility of a commodity, perhaps grown in a particular area, and whether the potential for contamination is apparent. Such information is derived from reviews of published literature, crop surveys for mold and/or mycotoxin contamination, and meteorological data.

The success of a monitoring program depends, in part, on establishing an appropriate sampling plan (Chapters 6 and 7) and using a validated analytical methodology (Chapter 9). Numerous commodity-specific sampling plans have been established for selected mycotoxins, and methods such as those validated by AOAC INTERNATIONAL are available. There should also be in place predetermined options for dealing with products that are found to be out of compliance, including alternative lower-risk uses or destruction.

OTHER CONTROL STRATEGIES

Where good agricultural practices have not been fully effective and mycotoxin formation has occurred, decontamination efforts are initiated to remove as much of the toxin from the contaminated material as technologically possible. Effective decontamination procedures can be broadly categorized into physical and chemical methods. Physical methods include electronic sorting, separation/segregation, and various forms of heat and solvent processing (14–16). Table 1.4 illustrates the effect of a combination of procedures on the final content of aflatoxin in a peanut product (17).

Table 1.4. Effectiveness of aflatoxin management strategies at the processing level for peanut products: same product treated serially through various processes

Technology	Residual aflatoxin (μg/kg)	Reduction (%)	Cumulative reduction (%)
Farmer's stock	217	—	—
Belt separator	140	35.5	35.5
Shelling operation	100	28.6	53.9
Color sorting	30	70.0	86.2
Gravity table	25	16.7	88.5
Glanching/color sorting	2.2	91.2	99.0
Recolor sorting	1.6	27.3	99.3

Chemical inactivation may take the form of structural modification/ degradation and adsorption. The ammoniation process degrades the aflatoxin molecule while chemoadsorption with phyllosilicates binds aflatoxin B_1 and restricts its absorption from the gastrointestinal tract (18, 19). The use of adsorbent materials and ammoniation has not been accepted by everyone. For instance, there is a belief that indiscriminate use of clays can pose a risk because essential nutrients may also be bound by some clays (20). The efficacy and safety of ammonia for reducing aflatoxin contamination has been amply demonstrated, hence its widespread use in Brazil, France, Senegal, the Sudan, and the southeastern U. S. for treating materials destined for animal feed (21). Individual governments will have to decide whether the process will be applied to human foods based on available scientific information and national laws. In the United States, the addition of chemical agents to human foods specifically for reducing mycotoxin levels is not permitted (22). When evaluating the effectiveness of toxin reduction processes, it is necessary to consider the chemical stability of the toxin, the potential interaction between the toxin and the food matrix, and the possible interaction between different toxins that may be present simultaneously.

INDUSTRY AND CONSUMER PARTICIPATION/ EDUCATION

Governments are responsible for protecting the public from involuntary hazards such as mycotoxins. Government agencies establish the standards and enforcement programs necessary to ensure that the public food supply is safe. These regulations govern how much of a potentially hazardous substance is permitted in specific foods and ensure compliance with the set limits in consumer-ready products; they do not directly affect mycotoxin contamination on the farm. Hence, efforts to prevent the formation of mycotoxins require the cooperation of farmers and processors. This cooperation depends on the effective dissemination of information about measures that can prevent mycotoxins from forming both before and after harvest and reduce levels during processing operations. Such an educational effort increases the awareness of safety concerns, identifies and assists in the implementation of preventive measures, and helps in the establishment of internal controls and operating standards. Consumer awareness of the potential for the occurrence of various toxins in food commodities and their effects on health is pivotal and can act as an impetus for legislating regulations by governments. Regulation will not succeed if the public is unaware of the risk or perceives such a risk as being of low priority.

GLOBAL HARMONIZATION OF MYCOTOXIN CONTROL EFFORTS

Several countries have developed specific regulations to control mycotoxin contamination in foods and feeds (23). In most countries, the setting of permissible levels is based, among other factors, on unique population considerations or the mandate of the country's societal level of public health protection. Risk managers in various countries address the issue with a combination of scientific judgment and socio-economic considerations. In the interest of fair international trade and universal protection of consumers, however, a call can be made to develop science-based internationally harmonized and recognized regulations to control mycotoxin occurrence in foodstuffs. Based on a scheme similar to that employed by the Joint Food and Agriculture Organization (FAO)/World Health Organizaion (WHO) Expert Committee on Food Additives (JECFA), it is feasible to establish maximum levels for mycotoxins in foods. Once mycotoxin limits are internationally recognized, they can be incorporated into the operating principles of existing global agreements, such as the World Trade Organization (WTO) Agreements on Sanitary and Phytosanitary Measures (SPS), and Technical Barriers to Trade (TBT), for greater harmonization and transparency of food regulations that protect consumers and facilitate trade.

SUMMARY

Contamination of food by naturally occurring toxins, including the mycotoxins, is unavoidable and unpredictable and poses a unique threat to food safety. Mycotoxins contaminate major agricultural commodities and animal-derived foods such as milk. Risks associated with mycotoxin-contaminated foods can be reduced through improved agricultural practices, regulatory limits, monitoring and decontamination. Several elements affect the risk assessment and management strategies of regulatory agencies that are establishing a food safety management program. Paramount are the expected gravity and/or probability of an adverse event occurring and the nature of the adverse event. Once the risk is assessed, an appropriate, economically sustainable management strategy is implemented. The success of the strategy is governed by science, effective risk communication, and consumer education. Regulation will not succeed if the public is unaware of the risk or perceives the risk as being of low priority. Although exposure to mycotoxins is undesirable, the limits established must not unduly compromise the availability of food or feed.

REFERENCES

(1)   *The Random House Dictionary of the English Language, 2nd Edition*, S.B. Flexner & L.c. Hauck (Eds), Random House, New York, New York, 1987

(2)   Lancaster, M.C., Jenkins, F.P., & Philip, J.M. (1961) *Nature* **192**, 1095

(3)   Smith, J.W., & Hamilton, P.B. (1970) *Poultry Sci.* **49**, 207

(4)   Krishnamachari, K.A.V.R., Ramesh, V., Bhat, V.N., & Tilak, T.B.G. (1975) *Indian J. Med. Res.* **63**, 1036

(5)   Ngindu, A., Johnson, B.K., Kenya, P.R., Ngira, J.A., Ocheng, D.M., Nandwa, H., & Omondi, T.N., Ngave, W., Gatei, B.K., Jansen, A.J., Kavite, J.N. and Siongok, T.A. (1982) *Lancet* **I**, 1346

(6)   Alma, I., Kamala, C.S., Gopalakrishna, G.S., Jayaral, A.P., Sreenivasamurthy, V., & Parpia, H.A.B. (1971) *Am. J. Clin. Nutr.* **24**, 609

(7)   FAO (1996) *Basic Facts of the World Cereal Situation, Food Outlook, No. 5/6* Food and Agriculture Organization, Rome, Italy

(8)   Boutrif, E., & Canet, C. (1998) *Rev. Med. Vet.* **149**, 681

(9)   Trujillo, S., Njapau, H., Park, D.L., Price, W.D., Pohland, A.E., Fremy, J-M., & Dragacci, S. (2001) in *GMO AND FOOD Can Benefits for Health be Evaluated,* Proceedings of the International Conference, December 17-18, 2001, Institut Pasteur, Paris, France, 78

(10)   Dowd, P.F. (2001) *J. Econ. Entomol.* **94**, 1067

(11)   Park, D.L., Njapau, H., & Boutrif, E. (1999) *FNA.* **23**, 49, Food and Agriculture Organization (FAO), Rome, Italy

(12)   Park, D.L., & Stoloff, L. (1989) *Regul. Toxicol. Pharm.* **9**, 109

(13)   CAST (2003) *Mycotoxins: Risks in Plant, Animal, and Human Systems,* Task Force Report No. 139, January 2003, Council for Agricultural Science and Technology, Ames, Iowa

(14)   Creppy, E.E. (2002) *Toxicol. Lett.* **127**, 19

(15)   FAO (1993) *Sampling Plans for Aflatoxin Analysis in Peanuts and Corn,* Food and Nutrition Paper 55, Food and Agriculture Organization, Rome, Italy

(16)   Phillips, T.D., Clement, B.A., & Park, D.L. (1994) in *The Toxicology of Aflatoxins—Human Health, Veterinary and Agricultural Significance*, D.L. Eaton & J.D. Groopman (Eds), Academic Press, San Diego, CA, 383

(17)   Park, D.L. (1993) *Food Technol.* **47**, 92

(18)   Park, D.L., Lee, L.S., Price, R.L., & Pohland, A.E. (1988) *J. Assoc. Off. Anal. Chem.* 71, 685

(19)   Park, D.L., & Liang, B. (1993) *Trends Food Sci. Tech .***4**, 334

(20)   Harvey, R.B., Kubena, L.F., & Phillips, T.D. (1993) *Sci Total Environ.* Suppl. 1993, 1453.

(21)   Park, D.L., & Price, W.D. (2001) *Rev. Environ. Contam. Toxicol.* **171**, 139

(22)   Phillips, T.D. (1995) *Nat. Toxins* **3**, 204

(23)   FAO (1997) *Worldwide Regulations for Mycotoxins 1995—A Compendium,* Food and Nutrition Paper 64, Food and Agriculture Organization, Rome, Italy

2. CONDITIONS LEADING TO THE OCCURRENCE OF MYCOTOXINS

Henry Njapau and Douglas L. Park

Center for Food Safety and Applied Nutrition
Food and Drug Administration
5100 Paint Branch Parkway
College Park, Maryland 20740

Earth is home to billions of living organisms, which are divided into five kingdoms. There are obvious and subtle interdependencies and interactions between the various inhabitants. The Kingdom Fungi encompasses a wide spectrum of largely multicellular organisms. Growth and development of fungi depend on physiological and biochemical processes that lead to the production of low molecular weight secondary metabolites. Although most are of no apparent benefit to the fungi, some, such as the trichothecenes may be essential for invasion of the substrate (Pitt, J.I., personal communication); a relatively small number may be beneficial (e.g. penicillin) or harmful to humans (e.g. mycotoxins). The mycotoxins, a classification that includes the trichothecenes, are fungal secondary metabolites that are associated with adverse health effects when consumed by humans and animals.

Although mycotoxins are classified as naturally occurring toxins, they are not natural products of the metabolism of corn, cotton or peanut plants. They are deposited in these food commodities by invading naturally occurring fungi. The naturally occurring classification of mycotoxins is based on the differentiation between anthropogenic toxicants ("man-made") and those toxins that are not a result of human activity. The terms toxin and toxicant have often been used interchangeably, the

13

latter term is associated with poisonous compounds of anthropogenic origin.

Mycotoxins are important because they contaminate a wide variety of agricultural commodities (Table 2.1) that are used for food and feed. When humans and animals consume food or feed contaminated with mycotoxins, they may be afflicted by a variety of diseases generally classified as mycotoxicoses. Mycotoxicoses are prevalent worldwide (1), however, only a small number of the known mycotoxins are associated with adverse health effects. For instance, a 1997 Plant Disease report by the University of Illinois Extension stated that, in the Midwestern United States, aflatoxin, zearalenone, and deoxynivalenol account for 99% of the diagnosed animal mycotoxicoses (2). Acute mycotoxin poisoning often results in the deterioration of liver function, which, in extreme cases, may lead to acute hepatitis and death. The effect of chronic exposure to low doses of the different toxins may vary. Interaction of the aflatoxins and other risk factors such as hepatitis B virus has been associated with a high prevalence of primary liver cancer in certain localities (3). The most prevalent mycotoxins (aflatoxins, fumonisins, deoxynivalenol, zearalenone, ochratoxin A and patulin) are produced by a small number of *Aspergillus*, *Fusarium*, and *Penicillium* species (1).

Table 2.1. Prevalent mycotoxins, producing fungi and most commonly affected agricultural crops

Mycotoxin	Fungi	Agricultural commodity	Reference
Aflatoxins	*A. flavus* *A. parasiticus*	Peanuts, corn, cottonseed, sorghum, spices, tree nuts	(1, 7, 25, 53)
Fumonisins	*F. verticillioides* *F. proliferatum*	Corn	(13, 14, 15, 27)
Deoxynivalenol	*F. graminearum*	Wheat, corn, barley, oats, rye	(15, 27)
Ochratoxin A	*A. ochraceaus* *P. verrucosum* *A. carbonarius*	Barley, wheat, grapes, coffee beans	(1, 11, 12)
Zearalenone	*F. graminearum*	Wheat corn, barley, oats, sorghum	(15, 27)
Ergot alkaloids	*Claviceps purpurea*	Rye, barley, millet	(1)

When mycotoxin contamination is suspected or ascertained, contaminated materials may be diverted to lower-risk uses or discarded. Recent estimates by the Food and Agriculture Organization (FAO) indicate that about 25% of the food crops worldwide are contaminated by mycotoxins every year (4). Such quantities cannot simply be discarded, particularly when almost 800 million people in developing countries do not have enough food to eat (5). In these localities, contaminated foodstuffs are probably consumed, perhaps leading to ill health. Therefore, it is important that mycotoxin contamination of food crops is reduced to minimal levels. The success of such efforts depends on how much we know and understand about what leads to the occurrence of these toxins in agricultural commodities. This chapter presents an overview of the factors that facilitate fungal infection and mycotoxin production in food crops.

FACTORS CONTRIBUTING TO MYCOTOXIN FORMATION

Although the precise biochemical events that trigger mycotoxin production in fungi are not well understood, the contributing external factors have generally been ascertained. Mycotoxins may be produced on the farm, and during transportation and storage of raw and processed foods as a result of interaction between the fungus, the food commodity, and the environment. A favorable combination of environmental factors facilitates infection and colonization of the food commodity by the fungus and influences the amount of mycotoxin produced.

Toxigenic Fungi

The most commonly occurring toxigenic species of *Aspergillus* Section Flavi (6), *Aspergillus flavus* and *Aspergillus parasiticus*, infect and produce aflatoxins in cereals, nuts and oilseeds (7). *Aspergillus flavus* produces aflatoxins B_1 and B_2 and the co-occurrence of aflatoxins B_1, B_2, G_1 and G_2 is attributed to *A. parasiticus* or *A. nomius* (8,9). Ochratoxin A is normally produced in some cereal grains by *Aspergillus ochraceus* in the tropics and by *Penicillium verrucosum* in temperate climates (1,10).

Recently, *Aspergillus carbonarius* has been identified as a major source of Ochratoxin A in coffee beans and grapes (11,12). The occurrence of patulin in apples and consequently apple juice is a result of infection by *Penicillium expansum* (10). *Fusarium* species are widespread and cause ear rot in wheat and corn (13,14). The most important toxigenic *Fusarium* species are *F. verticillioides* (synonym *F. moniliforme*), *F. proliferatum*, *F. graminearum* and *F. sporotrichioides* (15). *Fusarium verticillioides* produces fumonisins, *F. graminearum* produces deoxynivalenol (DON) and zearalenone (15). Before discovery of the fumonisins, *Fusarium* species were known to produce primarily the trichothecenes, zearalenone and moniliformin (1). Other fungi of toxicological importance include *Claviceps purpurea*, the producer of ergot alkaloids that were associated with a disease called St. Anthony's fire (16) in medieval times.

Susceptible Commodities

The world's major food and feed crops are excellent substrates for toxigenic fungi and can be infected as developing seed, in the later phases of growth, and in storage. However, their ability to support prolific fungal growth and toxin production varies. Peanuts, corn, and cottonseed readily support *A. flavus* growth, which results in the frequent occurrence of aflatoxins in these commodities. Conversely, less aflatoxin accumulates in soybean and rice, probably a result of multiple factors including chemical composition (17,18). Nevertheless, *A. flavus* has been isolated from peanuts, corn, cottonseed, rice, wheat, barley, rye, soybeans, coffee, pistachios, almonds, cashew nuts and figs (19,20). In corn profuse mycelial growth usually leads to discoloration of corn ears. *Aspergillus flavus* appears greenish-yellow whereas *F. verticillioides* infection usually appears white to salmon colored; a pinkish discoloration of corn kernels is typically associated with *F. graminearum* (15).

The Environment

The term environment is used here to represent the physical space within which a food crop is growing or stored. It includes the immediate vicinity within which an equilibrium may readily be attained as well as the wider ecological zone. At a particular temperature, the relative humidity of the air in the vicinity of the food commodity and the water in the food commodity attain an equilibrium due to moisture migration. This equilibrium, the equilibrium relative humidity (ERH), is directly correlated with an important food safety indice called water activity (a_w) (21, 22). Samapundo *et al.*, (23) have shown that *Aspergillus flavus* and *Aspergillus parasiticus* are unlikely to grow on corn below a water activity of 0.80 (approximately 13% moisture content).

In the subsequent sections of this chapter, moisture content expressed on a wet weight basis, is used as an indirect indicator of the potential water activity of a food commodity at a particular temperature and relative humidity because, moisture is more commonly used by grain handlers as a measure of potential grain spoilage than water activity. It should however, be understood that unlike water activity, the moisture content of a food does not explicitly portray the contribution of factors such as dissolved solutes which influence the growth of fungi and bacteria in the food commodity. For instance, differences in the susceptibility of dried fruits and cereals to fungal infection can readily be attributed to water activity. Dried fruits contain high levels of dissolved solids hence have a lower water activity than cereals at the same moisture content.

Worldwide, environmental conditions are generally suitable for fungal infection of agricultural commodities and proliferation, but extremes of temperature or reduced moisture limit growth. Moisture and temperature are therefore, the most critical factors that influence fungal infection of agricultural commodities on the farm or in storage and consequently the occurrence of mycotoxins. The most important toxigenic fungal species may grow on agricultural commodities in the temperature and moisture ranges of 2–48°C and 7–25%, respectively (10,24), and most require free oxygen for respiration. However, different fungi have different moisture and temperature requirements. *Aspergillus flavus* and *A. parasiticus* grow in the tropics whereas *Penicillium* species grow mostly in cooler climates. *Fusarium* species require higher moisture levels. Evidently, the environmental conditions that best support fungal growth are those that commonly occur in the world's major food-producing zones. Therefore, it is no wonder that mycotoxins have been isolated from most major food crops worldwide. Other important factors include drought, insect infestation, fungal spore load (inoculum) and sanitation

PREHARVEST MYCOTOXIN FORMATION

The farm is an important source of fungal contamination and subsequent mycotoxin formation. *Aspergillus* and *Fusarium* species can infect crops in the field when conditions are favorable. Often the favorable conditions that facilitate fungal invasion are a combination of difficult to control environmental factors and inappropriate agronomic practices.

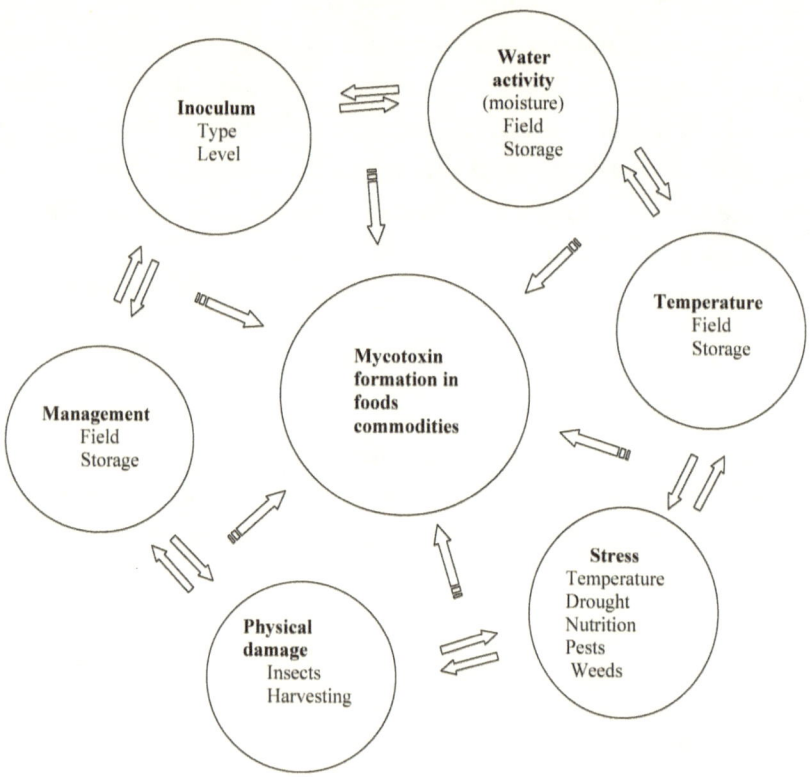

Figure 2.1. Interacting factors that contribute to the occurence of mycotoxins in agricultural commodities.

Preharvest infection of corn by *A. flavus* may occur through the silk of developing ears or through holes in the kernels resulting from insect infestation (25,26). In corn the fungus can grow at temperatures up to 45°C if the moisture content is within 14 to 20% (24,27,28). Drought, insufficient nutrients and crowding will stress the growing corn plant decreasing resistance to infection by fungi (29). While conditions suitable for fungal growth may not necessarily lead to the production of mycotoxins, aflatoxin formation by *A. flavus* in corn in the field may occur as the maturing kernels transition through the 20–13% moisture range when the temperature is 13–37°C, optimal production occurs between 17–18% moisture content and 25–32°C (30). Above a moisture content of 30%, *A. flavus* competes poorly with other fungi and bacteria hence it will normally not grow and produce aflatoxin in corn (31). On a peanut farm, soil-borne fungal spores are the primary source of infection of peanuts by *A. flavus*. The fungus may enter the developing peanut directly through the pod (32) particularly when the plant is stressed by high soil temperature and drought or the pods are damaged by insects (33). The biology and mechanism of infection of *Aspergillus parasiticus* have not been studied extensively but it has been reported (34,35) that the moisture and temperature requirements are similar to those of *A. flavus*. Pitt (9) however, noted that a substantial number of physiological studies reportedly conducted on *A. flavus* may have been carried out on *A. parasiticus*.

Fusarium mycotoxins are generally associated with cereals grown in temperate climates because toxigenic *Fusarium* sp. can grow and produce toxins at low temperature. However, occurrence of the fungi and their mycotoxins in food commodities has been documented globally (13,36). *Fusarium* sp. can thrive in decaying crop residue (37) that remain in the field after harvest allowing the fungus to sporulate during the subsequent growing season. The spores may be carried by wind and directly colonize new plants, or transmitted by insects from soil to plant and plant to plant (38). To infect and grow on standing corn *F. sporotrichioides* and/or *F. verticillioides* require a moisture content of about 18% (39) and minimum temperature of 2.0°C; both species will profusely grow between 22 and 27°C but not above 37°C (10). Production of the fumonisins on sterile corn by *F. verticillioides* and *F. proliferatum* has been demonstrated at 19% moisture and 20-30°C (40-42). These laboratory observations support the reported natural occurrence and distribution of fumonisins in tropical and subtropical environments (36,43).

Within an ecological zone, variations in meteorological conditions may affect different fungal species to various degrees. Under conditions of normal rainfall November to April), *Fusarium* species are the predominant field pathogens of corn in Zambia. When the rainfall season extends beyond "normal", the prevalence of *Diplodia macrospora* (*Stenocarpella macrospora*) supersedes that of *Fusarium* sp.; consequently, a decrease in *Fusarium* mycotoxins in one particular year may be substituted by an increase in the level of *D. macrospora* toxins, rendering the commodity still harmful to consumers.

The terminal on-farm operation is harvesting. During this activity, maturity of the crop, moisture and physical damage influence subsequent contamination. Crops harvested prematurely are more susceptible to insect damage and fungal infection. In some instances, a crop may mature and be harvested at a high moisture content. For example, it has been reported (44) that corn in the Mississippi delta reaches physiological maturity at about 30% moisture content. Depending on prevailing climatic conditions, the subsequent postharvest dry-down period may be sufficiently long for fungi to infect the corn and produce mycotoxins.

Grain damaged during mechanical harvesting is predisposed to rapid fungal infection, and, because mechanical harvesting is not selective, ears infected by fungi in the field and perhaps contaminated by mycotoxin may be included in the bulk going into storage. Where hand harvesting is the norm, visibly moldy ears are often discarded on the farm. Unless immediately destroyed the discarded ears may remain on the farm, increasing the soil spore load for the subsequent year (45).

POSTHARVEST MYCOTOXIN FORMATION

Throughout the world, storage fungi are predominantly species of *Aspergillus* and *Penicillium*. Mycotoxins that occur in storage may be a carryover from the field or a result of fresh fungal infection. The extent of fungal growth and toxin production in storage is governed by factors similar to those in the field, i.e. moisture, temperature, and biotic factors such as the presence of storage pests. Furthermore, the condition of the kernel, packaging, length of storage, and sanitary state of the storage facility will have an effect on the dynamics of fungal growth and toxin production. Storage fungi rapidly grow on commodities that are improperly dried or when the moisture of the stored product is allowed to increase. In un-insulated storage, diurnal temperature variations may cause moisture migration into the stored commodity. In warm and humid climates, the high relative humidity enhances moisture migration from the environment into stored grain. An increase in moisture during storage may also result from direct water infiltration from leaking warehouses and elevators. When high-moisture grain is kept in trucks, combines, or bins, for long periods of time, fungal and grain respiration elevate temperature in localized areas, which enhances mycotoxin production.

Fungal species that infect stored products require less moisture to grow and produce mycotoxins than field fungi. Nevertheless, the optimal temperature and moisture for toxin production in storage by a specific fungal species are similar to those under field conditions. The minimum moisture required for growth in cereals is about 14% for *Aspergillus* and 16% for *Penicillium* (46). At a given temperature, agricultural commodities in storage widely vary in susceptibility to fungal infection. It has been demonstrated (47) that while the safe storage moisture for corn at 20°C is 12%, it is 15% for wheat and soybean and about 7-8% for peanuts. *Aspergillus* species have very limited growth on corn when kernel moisture and temperature are below 12% and 10°C respectively (30). *Penicillium verrucosum* grows and produces ochratoxin A in wheat and barley in cool temperate zones from 0-31°C with an optimum near 20°C (48).

When optimal conditions are attained, *A. flavus* rapidly grows on corn or peanuts in storage. Aflatoxin production can start within 48 hours of initiation of rapid growth, and significant amounts of aflatoxin can be produced within 3-4 days.

Damage during harvesting and transportation and from elevated insect activity contribute to rapid deterioration of grain in storage. In addition to damaging the commodity, insects distribute spores throughout the stored product and can cause temperature and moisture to increase as a result of their metabolic activities. Insects have a high rate of reproduction (7) and penetrate a stack of bags far more quickly and thoroughly than bulk produce because of the gaps between the bags. Fumigation of large populations of insects may create a secondary substrate on which fungi can readily propagate, supplying greater quantities of inoculum (49).

Other inappropriate management practices that contribute to mycotoxin contamination include allowing stored commodities to come in contact with the ground (soil) and mixing clean and contaminated materials.

SUMMARY

Over the past 40 years, significant progress has been made in understanding the biology governing the interaction between fungi and agricultural commodities. Toxigenic species of the *Aspergillus*, *Fusarium* and *Penicillium* will produce mycotoxins when growing on food commodities under favorable conditions. Multiple physical, chemical, and biological factors, including poor management influence the degree of fungal infection and the formation of mycotoxins in agricultural commodities. Temperature, moisture, insect infestation, inoculum size, and the characteristics of the commodity are critical. It is evident that the meteorological conditions that support fungal growth and toxin production are similar to those required by agricultural commodities for optimal productivity. Consequently, total elimination of mycotoxin contamination has largely been unsuccessful. Success in managing mycotoxin contamination will, in the intervening period before "smart crops" are developed, depend on strategic management of the farm and storage environment in order to tip the balance toward sustaining food productivity while minimizing fungal growth and toxin production.

ACKNOWLEDGEMENT

The authors are grateful to John I. Pitt [CSIRO, Australia (retired)] and Valerie Tournas (U.S. Food and Drug Administration) for their valuable comments and suggestions.

REFERENCES

(1)   Wyllie, T.W., & Morehouse, L.G. (1977) *Mycotoxic Fungi, Mycotoxins, Mycotoxicoses, an Encylcopedic Handbook.* **1**, Marcel Dekker, New York

(2)   *Mycotoxins and Mycotoxicoses.* RPD No. 1105, (August 1997) University of Illinois Extension, Urbana-Champaign, IL

(3)   CAST (1989) *Mycotoxins: Economic and Health Risks, Task Force Report 116.* Council for Agricultural Science and Technology, Ames, IA

(4)   Boutrif, E., & Canet, C. (1998) *Rev. Méd. Vét.* **149**, 681

(5)   FAO (2002) *The State of Food Insecurity in the World 2002*, Food and Agricultural Organization, Rome, Italy

(6)   Gams, W., Christensen, M., Onions, A.H., & Pitt, J.I. (1985) in *Advances in Penicillium and Aspergillus Systematics*, R.A. Samson & J.I. Pitt (Eds), Plenum Press, New York, 55

(7)   Klich, M.A., & Pitt, J.I. (1988) *Trans. Br. Mycol. Soc.* **91**, 99

(8)   Hesseltine, C.W., Sorensen, W.G. & Smith, M. (1970) *Mycologia* **62**, 123

(9)   Pitt, J.I. (1993) *J. Food Prot.* **56**, 265

(10)   FAO (2001) *Manual on the Application of the HACCP System in Mycotoxin Prevention and Control*, Food and Nutrition Paper No. 73, Food and Agriculture Organization, Rome, Italy

(11)   Taniwaki, M.H., Pitt, J.I., Teixeira, A.A. & Iamanaka, B.T. (2003) *Int. J. Food Microbiol.* **82**, 173

(12)   Battilani, P., Pietri, A., Bertuzzi, T., Languasco, L., Giorni, P. & Kozakiewicz, Z (2003) *J. Food Prot.* **66**, 633

(13)   Placinta, C.M., D'Mello, J.P.F & Macdonald, A.M.C. (1999) *Anim. Feed Sci. Technol.* **78**, 21

(14)   Schaafsma, A.W. (1993) *Can. J. Plant Pathol.* **15**, 185

(15)   Marasas, W.F.O., Nelson, P.E., & Tuosson, T.A. (1984) *Toxigenic Fusarium Species: Identity and Mycotoxicology.* Pennsylvania State University Press, University Park, PA

(16)   Cadieux-Ledoux, H. (1985) *IDRC Reports.* **14**, 23.

(17)   Gardner, H.W., Grove, M.J., & Keller, N.P. (1998) *TEKTRAN*, U.S. Department of Agriculture, Agricultural Research Service, Peoria, IL

(18)   Pitt, J.I. & Hocking, A.D. (1994) *Int. J. Food Microbiol.* **23**, 35

(19)   Ellis, W.O., Smith, J.P., Simpson, B.K., & Oldham, J.H. (1991) *CRC Cr. Rev. Food Sci..* **30**, 403

(20)   Pitt, J.I., Hocking, A.D., Bhudhasamai, K., Miscamble, B.F., Wheeler, K.A. & Tanboon-Ek, P. (1993) *Int. J. Food Microbiol.* **20**, 221

(21)   Pitt, J.I. (1975) in *Water Relations of Food*, Duckworth, R.B. (Ed) Academic Press, London, 273

(22)   Labuza, T.P. and Hyman, C.R. (1998) *Trends Food Sci. Tech.* **9**, 47

(23)   Samapundo, S., Devlieghere, F., Geeraerd, A.H., De Meulenaer, B., Van Impe, J.F. & Debevere, J (2007) *Food Microbiol.* **24**, 517

(24)   Pitt, J.I. & Hocking, A.D. (1997) *Fungi and Food Spoilage, 2nd Ed,* Blackie, Academic & Professional, London, UK

(25)   Jones, R.K., Duncan, H.E., Payne, G.A. & Leonard, K.J. (1980) *Plant Dis.* **64**, 859

(26)   Munkvold, G.P., McGee, D.C., & Carlton, W.M. (1997) *Phytopathology* **87**, 209

(27)   Hill, R.A., Wilson, D.M., McMillian, W.W., Widstrom, N.W., Cole R.J., Sanders T.H., & Blankenship, P.D. (1984) in *Trichothecenes and Other Mycotoxins*, J. Lacey (Ed), Wiley and Sons, Chichester, UK, 79

(28)   Wrather, J.A., Francka, J.E., and Sweets, L.E., (2002) *Aflatoxin in corn*, Delta Research Center, Missouri Agricultural Experiment Station, University of Missouri, Columbia, MI

(29)   Lillehoj, E.B. (1983) in *Aflatoxin and Aspergillus flavus in Corn.* Southern Cooperative Service Bulletin 279, V.L. Diener, R.L. Asquith, & J.W. Dickens (Eds) Craftmaster, Opelika, AL

(30)   Brackett, R.E. (1989) in *Trends in Food Product Development*, C.T. Yam & C. Tan (Eds), Singapore Institute of Food Science & Technology, Singapore, 83

(31)   Oyebanji, A.O. & Efiuvwevwere, B.J.O. (1999) *Int. Biodeterior. Biodegrad.* **44**, 207

(32)   Griffin, G.J. (2001) *Soil Biol. Biochem.* **33**, 253

(33)   Bowen, K.L. & Mack, T.P. (1990) in *Highlights of Agricultural Research*, Alabama Agricultural Experiment Station, Winter 1990

(34)   Diener, U.L., Pettit, R.E. & Cole, R.J. (1982) in *Peanut Science and Technology*, H.E. Pattee & C.T. Young (Eds), American Peanut Research and Education Society, Yoakum, TX, 486

(35)   Pitt, J.I. & Miscamble, B.F. (1995) *J. Food Prot.* **58**, 86

(36)   Dutton, M.F. (1996) *Pharmacol. Therapeut.* **70**, 137

(37)   Stack, R. (1997) *Fusarium Head Bight of Wheat and Barley: An Overview*, *Fusarium* Scab Forum, November 10–12,1997, St. Paul, MN

(38)   Cardwell, K.F., Kling, J.G., Maziya-Dixon, B., & Bosque-Perez, N.A. (2000) *Phytopathology* **90**, 276

(39)   Gwimmer, J., Harnissch, R., & Muck, O. (1996) *Manual of the Prevention of Post-Harvest Grain Losses*. Post Harvest Project. Deutsche Gesellschaft fur Technische Zusammenarbeit, Eschbom, Germany

(40)   Alberts, J.F. (1990) *Appl. Environ. Microbiol.* **56**, 1729

(41)   Castella, G. (1999) *J. Food Prot.* **62**, 811

(42)   Marin, S., Homedes, V., Sanchis, V., Ramos, A.J. & Magan, N. (1999) *J. Stored Prod. Res.* **35**, 15

(43)   Sydenham, E.W. (1990) *J. Agric. Food Chem.* **38**, 1

(44)   Larson, E. (2002) *Minimizing Aflatoxin in Corn*, Mississippi State University Extension Service, MS

(45)   Wicklow, D.T., & Wilson, D.M. (1986) *T. Brit. Mycol. Soc.* **87**, 651

(46)   Sone, J. (2001) *J. Asia Pac. Entomol.* **4**, 17

(47)   *European Mycotoxin Awareness Network*, http://www.ifra.co.uk/ eman/fsheet3.htm.

(48)   Pitt, J.I. (2002) in *Mycotoxins and Food Safety: Advances in Experimental Biology and Medicine 504*, J.W. DeVries, M.W. Trucksess & L.S. Jackson (Eds), Kluwer Academic, New York, 29

(49)   Barney, R.J. (1995) *Crop Prot.* **14**, 159

(50)   Machinski, M. Jr., Soares, L.M.V., Sawazaki, E., Bolonhezi, D., Castro, J.L. & Bortolleto, N. (2001) *J. Sci. Food Agric.* **81**, 1001

Section II: Health Effects

3. HEALTH RISKS ASSOCIATED WITH MYCOTOXIN CONTAMINATION

Ronald T. Riley[1], Timothy Phillips[2], William P. Norred[1], Henry Huebner[2], and Shawna Lemke[2]

[1]Toxicology and Mycotoxin Research Unit
R. B. Russell Research Center
USDA Agricultural Research Service
P.O. 5677, Athens, Georgia 30604

[2]Intercollegiate Faculty of Toxicology
Veterinary Anatomy & Public Health
College of Veterinary Medicine
Texas A&M University
College Station, Texas 77843

The purpose of this review is to briefly summarize the current state of knowledge about the health risks associated with exposure to mycotoxins. Mycotoxins are toxic secondary fungal metabolites that cause disease if ingested in large enough quantities or over a long enough period of time. They are biochemicals produced by fungi from primary metabolites and secreted into the microenvironment occupied by the fungus and other microbes; when consumed or absorbed by animals, they can cause sickness and behavioral changes. It should be noted that this definition distinguishes between "fungal metabolites," "toxic fungal metabolites," and "mycotoxins."

WHY DO FUNGI MAKE MYCOTOXINS?

Although not necessary for growth of the fungus, secondary fungal metabolites may help the fungus by enhancing its ability to compete with other organisms, to invade host tissues, and to otherwise survive and reproduce. Nonetheless, the question "why do fungi produce mycotoxins?" has been asked by many but satisfactorily answered by none. A recent study suggests that toxigenic fungi may have secondarily originated from lichen-forming fungi that lost their lichen symbiotic habit (1). Lichen are a unique symbiosis between fungi and green algae or cyanobacteria. It was hypothesized that, when fungi assumed the lichen symbiosis, genes previously needed for primary metabolism were diverted to new functions. Conversely, loss of the lichen symbiosis may have resulted in non-lichen fungi that retained these "secondary" capabilities. Regardless of their origin, toxic fungal metabolites that can accumulate in foods and feeds vary considerably in their potency, in the symptoms they cause, in the target tissues they affect, and in their mechanisms of action. People have used molds since ancient times in the production of various foods, including cheese and salami, as well as in the fermentation of beer and wine. The secondary metabolites from many of these molds have been used as very effective antibiotics for the treatment of disease and as drugs for other important medicinal purposes.

HOW MANY MYCOTOXINS ARE KNOWN?

Mycotoxins that exist in our environment probably number in the thousands. They are produced by hundreds of thousands of species of fungi, many of which remain to be discovered (2). The variety in chemical structures of mycotoxins is readily apparent in Cole and Cox's *Handbook of Mycotoxins* (3). Nonetheless, the number of known mycotoxins that pose a measurable health risk to animals and humans is quite limited for several reasons. First, a basic tenet of toxicology is "the dose makes the poison." This means that, even though animals and humans are exposed to mycotoxins every day, the dose is insufficient for them to be acutely poisonous. Second, while there are thousands of publications that document the poisonous effects of toxic fungal metabolites in laboratory experiments *in vitro* and *in vivo*, the levels and routes of exposure do not model the exposure that occurs when farm animals and humans are exposed to naturally contaminated materials. Thus, we must be careful not to confuse the potential for toxicity with the documented and confirmed toxic effects in field situations. Nonetheless, the knowledge derived from *in vitro* studies and with laboratory animals serves as a red flag for the possible contribution of mycotoxins in altering immune function (4) and contributing to unexplained farm animal and human diseases as well as feed-associated performance problems in farm animals (5). Finally, for humans, cultural and socioeconomic conditions can place some groups at much higher risk than others. For example, subsistence farmers in underdeveloped countries consume larger amounts of potentially contaminated commodities just before they harvest the year's crops and when the food supply is not sufficiently diverse. Nutritional deficiencies can increase a person's susceptibility to mycotoxin-associated disease (as is the case with other diseases).

WHICH MYCOTOXINS ARE DANGEROUS?

For the purposes of this review, only those mycotoxins for which exposure is known to be high and those that are known to cause disease in animals and humans, are associated with human disease, or are suspected to be modifying factors in disease processes are presented. Mycotoxins with biochemical mechanisms of action that strongly suggest the action could modulate disease processes are also reviewed. On this basis, the mycotoxins that present the greatest risk to humans and farm animals are those that occur in commodities that are consumed in large amounts. Affected commodities include, for example, corn, wheat, sorghum, barley, millet, rye, peanuts, and to a much lesser extent rice. Mycotoxins in air-borne dusts, hay, and forage grasses are also considered. The mycotoxins of greatest concern are aflatoxins B_1 and M_1, cyclopiazonic acid (CPA), ochratoxin A, fumonisins B_1 and B_2, deoxynivalenol (DON), T-2 toxin, zearalenone, ergot alkaloids, ergot-like alkaloids, and macrocyclic trichothecenes. Other mycotoxins of lesser importance are also reviewed briefly.

This summary review draws heavily upon several documents that describe in great detail the health risks and mechanisms of action of economically important mycotoxins [for example, Council for Agricultural Science and Technology Task Force Report 138 (6) and the report of the Joint Food and Agriculture Organization (FAO)/World Health Organization (WHO) Expert Committee on Food Additives 56th Meeting (7)].

ASPERGILLUS TOXINS

Economically important mycotoxins produced by *Aspergillus* include aflatoxins (corn, peanuts, cottonseed, pistachios, walnuts, almonds, rice, copra, spices), CPA (corn, millet, peanuts), and ochratoxins (coffee beans, barley, wheat, maize, grapes).

Aflatoxins were first identified in the early 1960s as the cause of a mysterious outbreak in Great Britain, dubbed Turkey X disease. Despite the many implications of moldy food poisoning and disease, until the early 1960s, there was a lack of understanding about the consequences of mycotoxin exposure to human and animal health. A notable turning point came in the 1960s, when moldy feed was associated with the loss of thousands of young birds in Great Britain. Other mycotoxins may have been involved in this outbreak, most notably CPA. The outbreak killed over 100,000 turkeys, 14,000 ducklings, and thousands of partridge and pheasant poults that consumed imported feed that contained peanut meal heavily contaminated with *Aspergillus flavus* (8). An investigation revealed that the turkeys had been fed Brazilian peanut meal that contained four metabolic by-products-namely, aflatoxins B_1 (AFB_1), B_2 (AFB_2), G_1 (AFG_1) and G_2 (AFG_2) (9). While *A. flavus* produces only the B series, *A. parasiticus* produces aflatoxins of both the B and G series. The most troubling and also the most frequently encountered *Aspergillus* toxin is aflatoxin B_1. Aflatoxin B_1 was first found to be an extremely potent liver carcinogen in 1964 (10); shortly thereafter, liver cancer epizootics in farm-raised rainbow trout, which had been recurring since 1936, were found to be due to diets prepared with AFB_1- contaminated cottonseed (11). One of the major effects is a general reduction in weight gain for a variety of production animals, including pigs, cattle, and poultry (12).

Aflatoxins are fairly lipophilic molecules; therefore, absorption across the membranes of the gastrointestinal tract is the most likely mode of entry for the toxins into an organism. The compounds are also absorbed through the lungs (13). Once absorbed, the aflatoxins are distributed to most soft tissues and fat deposits. However, the major organs of accumulation are the liver and kidney, where biotransformation occurs (13). Phase I metabolism of the aflatoxins occurs primarily through cytochrome P_{450} enzymes and results in the hydroxylated products aflatoxin M_1 and aflatoxin B_2a, the O-demethylated product aflatoxin P_1, and the 8,9-epoxide. The toxicity of AFB_1 is attributed to its ability to undergo mixed function oxidation by cytochrome P_{450} to form a number of metabolites, some of which readily react with nucleic acids and proteins and cause cell death or tumor initiation.

Since their discovery, aflatoxins have been found to cause outbreaks of farm animal disease worldwide and have been implicated in human disease outbreaks in areas of the world where subsistence farmers consume contaminated crops. For example, a recent report attributed 12 deaths in Kenya's Meru North district to consumption of aflatoxin-contaminated corn (14). The primary effect linked to aflatoxin exposure in humans is liver cancer. Although the incidence of this disease is low in the United States (0.5/100,000 deaths per year), in parts of Africa and Asia the numbers can reach 3000/100,000 deaths per year (15). Epidemiological studies show a positive correlation between intake of aflatoxin and liver cancer among African and Asian populations. For example, the relative risk for liver cancer associated with consuming a large amount of aflatoxin and a minor amount of alcohol has been shown to be 17.5, compared with a relative risk of 3.9 when alcohol consumption was heavy and aflatoxin intake was light (16). Some species are much more sensitive to the liver toxicity and tumorigenicity of AFB_1. Risk factors that contribute to an individual's sensitivity to aflatoxin-induced liver tumors include level of exposure to aflatoxin, expression of aflatoxin activation/detoxification pathways, nutritional status, and, most importantly, chronic infection with hepatitis B or C virus (17). A diet dependent on foods such as cassava, peanuts, sweet potato, and corn increases the likelihood of consuming mold-infected food. Aflatoxin B_1 induction of liver tumors in laboratory animals is closely correlated with mutations to specific genes that are known to control tumorigenicity. The same genetic alterations have been found in humans who consume large amounts of AFB_1-contaminated food, particularly if exposure to hepatitis B or C virus has occurred (17). It has been shown that infection with hepatitis B or C virus acts synergistically with AFB_1 to induce mutations known to cause tumors. The nature of mutations in rats caused by aflatoxin exposure and the incidence of similar mutations in humans with hepatocellular carcinoma implicate this mycotoxin as a human carcinogen. The best strategy for reducing aflatoxin-induced liver cancer risk is through implementation of a universal hepatitis B virus vaccination program in high-risk areas (17).

In addition to the fact that AFB_1 could be an important contributor to liver disease in adults, aflatoxin M_1 (AFM_1), a metabolite of AFB_1, is found in milk of dairy cows and other animals that consume aflatoxin-contaminated feeds (7). Also, milk production in dairy cattle is decreased in the presence of contaminated feed (18), and the milk carries AFM_1. This compound, although less toxic than AFB_1 (7), is regulated in the United States (0.5 µg/kg in milk) because of the particular susceptibility of the young, who consume large quantities of milk (13). Aflatoxins have also been implicated in Kwashiorkor disease and Reye's syndrome (18,19). Although the evidence for aflatoxin causing these diseases is not strong, the possible involvement of aflatoxin cannot be ruled out because higher levels of aflatoxin in tissues of affected individuals and hepatic involvement is apparent (6, 19).

CPA is produced by several fungi; however, in corn and peanuts, the primary producers are *Aspergillus* species, including *A. flavus*. In certain mold-ripened cheeses and meats, several *Penicillium* species produce CPA. Although CPA is not as toxic as aflatoxin, and apparently not carcinogenic, concern about its presence in foods stems primarily from its co-occurrence with AFB_1 in peanuts and corn and its mechanism of action, which is similar to that of the known tumor promoter thapsigargin (20). Cyclopiazonic acid has been implicated as a causative agent in several cases of field intoxications, including in pigs, cattle, quail, and humans (21). Acute doses of CPA administered to rats cause toxic lesions in liver, spleen, gastrointestinal tract, and skeletal muscle (22). Cyclopiazonic acid is a potent inhibitor of Ca^{2+} uptake and Ca^{2+}-dependent ATPase activity in both sarcoplasmic and endoplasmic reticulum (20). Although CPA is a useful biochemical tool for studying sarcoplasmic and endoplasmic reticulum function, there is little evidence to implicate CPA in human disease (19).

Ochratoxins, like CPA, are produced by several fungi, including several species of *Aspergillus*. In cool temperate areas, ochratoxins are also commonly produced by *Penicillium verrucosum*. There is little doubt that ochratoxins cause farm animal disease, most notably porcine nephropathy in northern Europe (23). Since the 1950s, the eastern European countries Bulgaria, Romania, and Yugoslavia have displayed a high rate of kidney disease known as Balkan endemic nephropathy. The prevalence of the disease has reached as high as 8.3% of the population in recent years (24). Ochratoxin A has been shown to induce proximal tubular dysfunction in many animals and is a renal carcinogen in mice (6). The mechanism of action of ochratoxin is unclear; however, its structural similarity to phenylalanine and the fact that it inhibits many enzymes and processes that are phenylalanine dependent strongly suggest that ochratoxin A acts by disrupting phenylalanine metabolism (25). Whether ochratoxin is the cause of any human disease, including Balkan endemic nephropathy, is much debated (7, 26). Nonetheless, the presence of high levels of ochratoxin A in human blood in both Europe and Africa (27, 28), coupled with the fact that it is a known cause of renal disease in farm animals and of liver tumors in rodents, is a clear reason for concern.

Other mycotoxins produced by *Aspergillus* species and found on cereal grains and other commodities likely to be consumed by humans and farm animals include sterigmatocystin, "tremorgens," and patulin. Sterigmatocystin is a precursor of aflatoxin and, like AFB_1, it is a liver carcinogen. Tremorgens are defined as fungal metabolites that elicit tremors in animals (3). They are produced by several genera of fungi and those produced by *Aspergillus fumigatus* have been implicated in farm animal diseases associated with improperly handled corn silage (6). Patulin is produced by *Aspergillus* and *Penicillium* species and other fungi. Its health risks are described in the section on *Penicillium* toxins.

FUSARIUM TOXINS

Economically important mycotoxins produced by *Fusarium* include fumonisins (corn and, to a much lesser extent, rice), DON and T-2 toxin (corn, wheat, barley, oats, millet, and, to a lesser extent, rice, rye, soybeans, and beans), and zearalenone (corn, wheat, oats, barley, rye, sorghum, soybeans, rice). The first record of health effects from consuming grain putatively contaminated with trichothecene mycotoxins was noted in 1913 in Russia (29). Consumption of over-wintered wheat produced symptoms of oral inflammation and vomiting followed by extensive hemorrhages in the skin and necrotic ulcers around the mouth, a syndrome now known as alimentary toxic aleukia.

Fumonisin B_1 is the most prevalent member of a family of mycotoxins produced by several species of *Fusarium*. Unlike most of the mycotoxins reviewed here, fumonisins are highly polar and water soluble. The fumonisins were recently the subject of comprehensive reviews by the WHO International Program on Chemical Safety (30) and the 56th Joint Expert Committee on Food Additives (7).

Fumonisins are causative agents of two farm animal diseases: equine leukoencephalomalacia (ELEM) and porcine pulmonary edema (PPE) (30). Equine leukoencephalomalacia is a fatal disease that has been described in horses and donkeys and was linked to moldy corn as early as the 1930s (30). Affected animals develop a lack of appetite, followed by a period of lethargy. In the final stages of the disease, the animal shows signs of uncoordinated movement and becomes ill tempered (30). The animal may show additional neurological irregularities including facial paralysis and seizures. Horses appear to be the animals most sensitive to fumonisin ingestion and the only species that develop brain lesions in response to the toxin. Porcine pulmonary edema is a fatal disease that develops in pigs and is characterized by pulmonary edema, hydrothorax, cardiovascular dysfunction, and liver toxicity (30).

In addition to the species-specific ELEM and PPE, fumonisins cause toxic effects that are common in most if not all species. The most common target organs of all species studied are liver and kidney. Fumonisin B_1 is a non-genotoxic renal carcinogen in male rats, a hepatocarcinogen in female mice (31), and a promoter of AFB_1 liver tumors in rainbow trout (32). A review of the effects of fumonisins in poultry suggested that, although the toxins are capable of producing disease and death, they are not a serious threat because a relatively high dose (75 mg/kg in the feed) is needed to produce a minimal effect (13). This dose is well above what has been reported as average feed contamination. However, at higher levels, fumonisin B_1 and fumonisin culture material produce decreased body weight gain and an increase in the weights of the liver, proventriculus, and gizzard of chickens.

The fumonisins bear a remarkable resemblance to sphinganine, a building block of sphingolipids needed for the structure and function of cellular membranes (30). There is considerable evidence that the underlying mechanism by which fumonisins cause toxicity to animals is disruption of lipid metabolism. Most notably, fumonisins are specific inhibitors of ceramide synthase (sphinganine and sphingosine N-acetyltransferase), a key enzyme in the pathway leading to formation of ceramide and more complex sphingolipids. Equine leukoencephalomalacia, PPE, and hepato- and nephrotoxicity that result from fumonisin exposure are closely correlated with the degree of disruption of sphingolipid metabolism (7, 30). Elevation of free sphingoid bases in serum and urine has been proposed as a biomarker for exposure to fumonisins. This biomarker works well in farm animals and has had mixed success in humans (33–35). A recent study has found that fumonisins can be detected in human hair (36).

Fumonisins have been associated with esophageal cancer, reproductive toxicity (neural tube defects), and acute disease outbreaks among humans who consume poor-quality corn as a staple (7,30). Epidemiological evidence correlates high rates of this form of cancer in the Transkei region of South Africa with consumption of corn with high levels of contamination of *Fusarium moniliforme*. In all cases, there is only limited evidence for fumonisins being the cause of these diseases in humans (7). Nonetheless, exposure in some developing countries at levels that can exceed the recommended provisional maximum tolerable daily intake (7), its co-occurrence with AFB_1 (and possibly ochratoxin A), and its unusual mechanism of action make it a mycotoxin of considerable concern in countries where poor-quality corn can make up a significant portion of the total caloric intake (30).

The trichothecenes are a large group of structurally related mycotoxins produced by *Fusarium*, *Myrothecium*, *Trichoderma*, *Stachybotrys*, and probably other fungal species. General signs of trichothecene toxicity in animals include weight loss, decreased feed conversion, feed refusal, low leukocyte counts, vomiting, bloody diarrhea, severe dermatitis, hemorrhage, oral lesions, decreased egg production, spontaneous abortion, and death (18). With regard to farm animal and human exposure through ingestion of contaminated feeds and foods, DON and T-2 toxin produced by *Fusarium* species are by far the most important. Deoxynivalenol is of special importance because of (i) its prevalence as a contaminant of wheat, corn, and barley; (ii) its frequent occurrence in farm animal performance problems, which results in severe economic losses; and (iii) the possible involvement of DON in modulating immune response and kidney toxicity (4). T-2 toxin (and to a somewhat lesser extent diacetoxyscirpenol) is also an important concern, because the mechanism of action is the same as for DON, but it is much more toxic and has also been implicated as a cause of farm animal and human disease (7, 19). Skin contact with T-2 toxin can cause dermal necrosis. Southeast Asians exposed to a suspected chemical warfare agent known as "yellow rain" presented similar symptoms, which suggests that T-2 may have been an active component (37).

Farm animal performance problems are a common manifestation of DON exposure. Swine refuse feed that contains DON at relatively high concentrations; at lower concentrations, DON can cause profuse vomiting. For this reason, DON is also referred to as vomitoxin. Like other trichothecenes, DON is an inhibitor of protein synthesis; it binds to ribosomes and inhibits translation. It is also a potent modulator of immune response in mice (4). In mice, DON causes a renal disease known as IgA nephropathy, which may be related to its ability to cause superinduction of proinflammatory cytokines (4). The manifestations of IgA nephropathy in mice are very similar to a human disease of the same name. However, there is no evidence that DON contributes to renal disease in humans. There have been several reports of human disease outbreaks associated with the consumption of moldy cereals or "scabby" wheat contaminated with T-2 toxin, DON, and other *Fusarium* mycotoxins. Typically, the manifestations involve gastrointestinal problems and, in the case of Alimentary Toxic Aleukia, degeneration of blood-forming tissues and other blood abnormalities (19). Although the data for involvement of DON and T-2 toxin in these disease outbreaks are supportive, they are not conclusive (19). Nonetheless, there is good reason to believe, based on *in vitro* and animal toxicity studies (4) and the reported population-based and ecological studies in humans, that consumption of foods highly contaminated with the trichothecenes DON and T-2 toxin will cause acute poisoning in humans (7).

Zearalenone is an estrogenic mycotoxin that often co-occurs with DON on scabby wheat and on corn that shows signs of *Fusarium* (*Gibberella*) ear rot. Both zearalenone and DON are produced by *Fusarium graminearum*. There is no doubt that zearalenone causes farm animal performance problems. It has been shown to cause vulvovaginitis and estrogenic responses in adult pigs and estrogenism in suckling piglets exposed during lactation via the sow's milk (6). Zearalenone produces dramatic effects in swine reproduction, including precocious sexual development in gilts, vulva enlargement, anestrus, pseudopregnancy, reduced litter size and reduced viability of newborn pigs (38). In boars, zearalenone can cause a reduction in the size of testes, epididymis, and vesicular gland weights as well as cessation of spermatogenesis and reduced libido. Other animals have shown less dramatic effects in response to zearalenone. Zearalenone acts by binding to the estradiol receptor; however, it binds with less affinity and is much less potent than estradiol (5).

Zearalenol, the hydroxy metabolite of zearalenone, has anabolic characteristics and is used to increase weight gain in cattle, sheep, and swine (18). Zearalenone and zearalenol used as growth promoters in swine were suggested to be involved in precocious development in children in Puerto Rico and other countries; however, the evidence is inconclusive (6). Climatic conditions favoring fungal growth and production of *Fusarium* toxins, including zearalenone, have been hypothesized to be risk factors for hormone-dependent cancer and adverse reproductive outcomes in farmers' families (39). However, the study did not provide any data about exposure to any *Fusarium* toxin.

Other mycotoxins produced by Fusarium species and found on cereal grains (primarily corn) and other crops likely to be consumed by humans or farm animals include nivalenol, fusarenon X, fusarochromanone, moniliformin, fusaric acid, and fusarin C. These mycotoxins probably are of little or no risk to consumers of cereal grains. Fusarochromanone has been suggested to play a role in a disease of poultry called tibial dyschondroplasia. Fusarin C, which can co-occur with fumonisins, is genotoxic and has been suggested to be involved in the rodent carcinogenicity of *Fusarium verticillioides* (synonym = *F. moniliforme*) culture material. Wortmannin is a *Fusarium* toxin that has been suggested as a possible agent in onyalai disease in south-central Africa, a human disease of unknown origin characterized by hemorrhaging in the mouth, tongue, palate, skin, and gastrointestinal tract (19).

PENICILLIUM TOXINS

Economically important mycotoxins produced by *Penicillium* include ochratoxins, CPA (described in the *Aspergillus* toxins section), and patulin (fruit juices). Of lesser importance are rubratoxin (corn), citreoviridin (rice), citrinin (moldy feeds), and penitrem A (moldy foods). Compared with the toxins produced by *Aspergillus* and *Fusarium*, the farm animal and human health risks from *Penicillium* toxins in foods (with the exception of ochratoxin A) are much more difficult to document. Patulin is of some concern because it is often found in apple juice, where it has been reported to reach concentrations as high has 45 mg/L (40); however, concentrations between 1 µg/L and 1 mg/L are the norm. There is no evidence that apple products contaminated with patulin are a cause of farm animal or human disease (40). Citreoviridin is now believed to have been the cause of acute cardiac beri-beri in Japan (yellow rice toxic syndrome); however, this disease has not been seen since laws were passed to prevent moldy rice from entering the marketplace in Japan (6). Rubratoxin and citrinin are suspected but not proven of having caused or contributed to farm animal diseases. Penitrem A has been reported to cause sickness in domestic animals and humans (6).

CLAVICEPS TOXINS

Ergotism is the name given to the acute diseases associated with consumption of cereals (primarily rye, barley, and millet) that are contaminated with sclerotia (ergot) from several *Claviceps* species. The ergot (sclerotia) can contain large amounts of clavine alkaloids, lysergic acids, lysergic acid amides, and peptide alkaloids (41). If the ergot is not removed from the cereal grain, then humans who consume the contaminated cereal will exhibit symptoms of the acute diseases known as ergotism (12). Ergot poisoning from feeds and forage grasses infected with *Claviceps purpurea* is also seen in pigs, cattle, and other farm animals (5). The acute toxic effects of ergot- and ergot-like poisoning in farm animals are similar and often involve central nervous system effects and gangrenous effects that result from the potent vasoconstrictive activity of some of the alkaloids. Other indications in livestock are numerous and may involve increases in body temperature; convulsions; muscle spasms; reduction in weight gain, conception, and milk production; and decreased serum prolactin levels. Neurological effects in animals may include ataxia, convulsions, and paralysis; chronic exposure to ergot alkaloids often results in redness, swelling, and eventual necrosis of the extremities. Lameness and necrosis of ears, tails, and feet have also occurred in animals pastured on endophyte-infected grasses (42). In swine, ergot-contaminated diets have been associated with increased stillborn rates, decreased piglet birth weights, and altered gestation times (43). Tall fescue and perennial ryegrass can also be endophytically colonized with *Neotyphodium* species that produce ergot-like alkaloids (6).

With regard to human health, ergotism is thought to be the oldest recognized form of mycotoxicosis. This disease became epidemic in the Middle Ages, when people consumed foods made with ergot-contaminated grains. After exposure, people had burning sensations, swollen limbs, and subsequent necrosis and loss of appendages (44). The most recent outbreaks of ergotism in humans occurred in the mid to late 1970s (19). More than 90 cases of human gangrenous ergotism have been reported in Africa, where people consumed grain infected with *Claviceps purpurea* (45). In addition, 78 cases of ergotism involving millet infected with *Claviceps* sp. have been reported in India (46). Other manifestations of ergotism included symptoms such as tingling under the skin, pruritis, numbness of extremities, muscle cramps, convulsions, and hallucinations (44). These symptoms primarily were due to the neurotoxic and vasoconstrictive actions of compounds that belong to the peptide alkaloid group (47).

In humans, ergotamine is metabolized by the liver largely via undefined pathways, and 90% of the metabolites are excreted in the bile. The unmetabolized compound is erratically secreted in the saliva, and only traces of unmetabolized ergotamine appear in the urine and feces. The elimination half-life of ergotamine from plasma is about 2 hours; however, ergotamine may be stored in some tissues, which may account for its long-lasting actions (48). Importantly, when the biological mechanisms of ergot alkaloid toxicity and potency are being determined, the following factors must be considered: (i) these alkaloids interact with more than one type of specific receptor site, (ii) the nature of receptor sites to which the alkaloids have access varies from organ to organ, and (iii) the affinity and efficacy of toxin and receptor vary greatly and are a function of the chemical configuration of alkaloids (49).

STACHYBOTRYS TOXINS

Stachybotrys toxins are currently the focus of considerable interest because they have been associated with what is called sick building syndrome (50), which has many similarities to the animal disease stachybotryotoxicosis that was originally described as a fatal disorder of horses, sheep, and goats and purportedly was widespread in Eastern Europe in the 1930s (51). Sick building syndrome involves respiratory illness in people inhabiting buildings that have been water damaged and where there is extensive fungal contamination (52). There is considerable debate about whether mycotoxins are involved in these human illnesses (53). In farm animals, the causative organism for stachybotryotoxicosis was identified as *Stachybotrys alternans* (*Stachybotrys atra* Bisby) by Russian scientists. Several species of *Stachybotrys* are known to also cause stachybotryotoxicosis in humans exposed to *Stachybotrys*-contaminated feeds (6). The causative toxins are believed to be macrocyclic trichothecenes, which are probably the most acutely toxic of all the mycotoxins. Like T-2 toxin, they are potent immunomodulators and inhibitors of protein synthesis (4).

MISCELLANEOUS TOXINS

Other mycotoxins that are known or suspected of causing farm animal and human diseases include sporidesmin, slaframine, and toxins produced by *Alternaria* sp. Sporidesmin is produced by *Pithomyces chartarum* and is found in dead and decaying plants in pastures (54). Consumption of sporidesmin causes liver toxicity in many grazing animals and intoxication is characterized by sunlight-induced "eczema" in light-skinned farm animals (typically in the facial region). Slaframine is a mycotoxin found in red clover and produced by *Rhizoctonia leguminicola*. It causes "slobbers" or excess salivation in cattle and other farm animals (6). Toxins produced by *Alternaria* on cereal grains (barley) have been implicated in contributing to Kashin–Beck disease in Tibet, a multifactorial disease characterized by joint deformities (55). Kashin–Beck disease has also been attributed to the *Fusarium* toxin fusarochromanone (19).

IS THE RISK FROM MYCOTOXINS UNDERESTIMATED?

This review has focused on "economically important" mycotoxins. Economically important means that the mycotoxin can be found in marketed commodities at levels approaching those that can cause biochemical or cellular effects, and/or there is some evidence that the mycotoxin has caused a farm animal or human disease outbreak. Unfortunately, new economically important mycotoxins are usually discovered as the result of a farm animal or human disease, which at the time of the first outbreak is of unknown etiology (e.g., aflatoxins and fumonisins).

The economic and health effects of known mycotoxins are certainly important. However, knowing the actual occurrence and distribution of all potentially toxic chemicals in foods and feeds is also important; because exposure to unknown bioactive agents could be an important confounding factor in the attempt to explain the etiology of chronic disease in animals and humans. To understand the health risk from mycotoxins in our environment, it is necessary to have a good estimate of mycotoxin exposure. To make these estimates we need to know all the mycotoxins to which we are exposed and we need to have some estimates of the doses. This is not possible with our current state of knowledge, because many mycotoxins have yet to be discovered. The precise number of secondary metabolites or mycotoxins is uncertain. A rough estimate of the number of yet to be discovered mycotoxins has been made by using several very debatable assumptions (56).

A calculation (56) is presented here to make the point that our ignorance about the occurrence and biological activity of toxic fungal metabolites could lead us to grossly underestimate farm animal and human exposure to mycotoxins. Novel disease outbreaks of unknown cause undoubtedly will lead to economic losses. It is possible to roughly estimate the number of fungal metabolites and potential mycotoxins (56). In 1971, Turner (57) catalogued about 1,200 secondary fungal metabolites produced by about 500 species of fungi. In 1983, Turner and Alderidge (58) catalogued 2000 more metabolites produced by about 1,100 species. Thus, as an average estimate, there were approximately two unique secondary metabolites per fungal species. In 1991, Hawksworth (2) estimated that there were 69,000 known fungal species and estimated that this represented 5% of the total species in the world, which Hawksworth estimated at 1,500,000. Conservative estimates are in the order of 100,000 species (59). Based on the work of Hawksworth (2) and the assumption of two unique secondary metabolites per fungal species, there may be as many as 3,000,000 unique secondary fungal metabolites. A conservative estimate would be 200,000. Between 1971 and 1983, the total number of known secondary fungal metabolites increased from 1,200 (57) to 3,200 (58). Assuming that the rate of discovery (accessible in the published literature) remained at a similar rate, there would be about 6,900 by the year 2005. This is 0.2% of the 3,000,000 or 3% of the 200,000 total estimated above. Clearly, the number of undiscovered secondary metabolites is quite large. Cole and Cox (3) listed about 300 secondary fungal metabolites as mycotoxins. About 10% of the secondary fungal metabolites described by Turner (57) and by Turner and Alderidge (58) were classified as mycotoxins by Cole and Cox (3). Thus, there are potentially between 20,000 and 300,000 unique mycotoxins.

SUMMARY

This review has briefly summarized the health risks associated with exposure to mycotoxins. Mycotoxins vary considerably in their potency, in the symptoms they cause, in the target tissues affected, and in their mechanisms of action. Mycotoxins probably number in the thousands, and they are produced by hundreds of thousands of species of fungi. Nonetheless, the known mycotoxins that pose a measurable health risk to animals and humans are quite limited. When dealing with food-borne chronic diseases it is very difficult to establish causation because, unlike acute diseases, the onset of symptoms seldom occurs concurrent with exposure. Most mycotoxin-associated diseases are probably multifactorial and, in many cases, mycotoxins are not prime causes but they contribute to increased susceptibility to diseases. The fact that mycotoxins are often suspected in farm animal disease outbreaks and production problems in developed countries is a warning of the potential of mycotoxins to contribute to diseases of unknown etiology in humans who consume poor-quality commodities in large amounts.

REFERENCES

(1)   Lutzoni, F., Pagel, M., & Reeb V. (2001) *Nature* **411**, 937

(2)   Hawksworth, D.L. (1991) *Mycol. Res.* **95**, 641

(3)   Cole, R.J., & Cox, R.H. (1981) *Handbook of Toxic Fungal Metabolites*, Academic Press, New York, NY

(4)   Bondy, G.S., & Pestka, J.J. (2000) *J. Toxicol. Environ. Health B* **3**, 109

(5)   Osweiler, G.D. (2000) *Vet. Clin. North Am. Food Anim. Pract.* **16**, 511

(6)   CAST (2001) *Mycotoxins: Risks in Plant and Animal Systems. Task Force Report 138*, Council for Agricultural Science and Technology, Ames, IA

(7)   WHO (2001) *Safety Evaluation of Certain Mycotoxins in Food*, WHO Food Additives Series, No. 47, World Health Organization, Geneva, Switzerland

(8)   Blount, W.P (1961) *Turkeys* **9**, 52

(9)   Asao, T., Buchi, G., Abdel_Kader, M. M., Chang, S. B., Wick, E. L., & Wogan, G. N. (1963) *J. Am. Chem. Soc.* **85**, 1706

(10)   Butler, W.H., & Barnes, J.M. (1964) *Nature* **201**, 1016

(11)   Jackson, E.W., Wolf, H., & Sinnhuber, R.O. (1968) *Cancer Res.* **28**, 987

(12)   Smith, J.E., & Ross, K. (1991) in *Mycotoxins and Animals Foods*, J.E Smith & R.S Henderson (Eds), CRC Press, Boca Raton, FL, 101

(13)   Leeson, S., Diaz, G.J., & Summers, J.D. (1995) *Poultry Metabolic Disorders and Mycotoxins*, University Books, Guelph, Ontario, Canada

(14)   *The Nation* (2001), Oct 3, 2001, Nairobi, Kenya

(15)   Groopman, J.D., Cain, L.G., & Kensler, T.W. (1988) *CRC Cr. Rev. Toxicol.* **19**, 113

(16)   Bulatao-Jayme, J., Almero, E.M., Castro, C.A., Jardeleza, T.R., & Salamat, L. (1982) *Int. J. Epidemiol.* **11**, 112

(17)   Henry, S., Bosch, F.X., Troxell, T.C., & Bolger, P.M. (1999) *Science* **286**, 2453

(18)   CAST (1989) *Mycotoxins: Economic and Health Risks*, Task Force Report No. 116, Council for Agricultural Science and Technology, Ames, IA

(19)   Wild, C.P., & Hall, A.J. (1996) in *The Mycota VI. Human and Animal Relationship*, D.H. Howard & J.D. Miller (Eds), Springer-Verlag, Berlin, Germany, 213

(20)   Riley, R.T., Goeger, D.E., & Norred, W.P. (1995) in *Molecular Approaches to Food Safety*, M. Eklund, J.L. Richard & K. Mise (Eds), Alaken Inc., Fort Collins, CO, 461

(21)   Bryden, W.L.(1991) in *Emerging Food Safety Problem Resulting from Microbial Contamination*, K. Mise & J.L. Richard (Eds), Ministry of Health and Welfare, Tokyo, Japan, 127

(22)   Norred, W.P., & Riley, R.T. (2001) in *Mycotoxins and Phycotoxins in Perspective at the Turn of the New Millennium*, W.J de Koe, R.A Samson, H.P. van Egmond, J. Gilbert & M. Sabino (Eds), W.J. de Koe, Hazekamp 2, Wageningen, The Netherlands, 211

(23)   Fink-Gremmels, J. (1999) *Vet. Q.* **21**, 115

(24)   Ceovic, S., Plestina, R., Miletic-Medved, M., Stavljenic, A., Mitar, J., & Vukelic, M. (1991) in *Mycotoxins, Endemic Nephropathy and Urinary Tract Tumors*, M. Castegnaro, R. Plestina, G. Dirheimer, I.N. Chernozemsky & H. Bartsch (Eds), International Agency for Research on Cancer (IARC), Lyon, France, 5

(25)   Riley, R.T., & Norred, W.P. (1996) in *The Mycota VI. Human and Animal Relationships.* D.H. Howard, & J.D. Miller (Eds), Springer-Verlag, Berlin, Germany, 193

(26)   Tatu, C.A., Orem, W.H., Finkelman, R.B., & Feder, G.L. (1998) *Environ. Health Perspect.* **106**, 689

(27)   Peraica, M., Radic, B., Lucic, A., & Pavlovic, M. (1999) *Bull. World Health Organ.* **77**, 754

(28)   Jonsyn-Ellis, F.E. (2001) *Mycopathologia* **152**, 35

(29)   Smith, J.E., & Moss, M.O. (1985) in *Mycotoxins: Formation, Analysis and Significance*, J.E. Smith & M.O. Moss (Eds), Wiley, New York, NY, 73

(30)   WHO (2000) *Environmental Health Criteria 219: Fumonisin B₁, EHC 219*, W.F.O. Marasas, J.D. Miller, R.T. Riley & A. Visconti (Eds), International Programme on Chemical Safety, United Nations Environmental Programme, The International Labour Organization, and the World Health Organization, Geneva, Switzerland

(31)   Howard, P.C., Eppley, R.M., Stack, M.E., Warbritton, A., Voss, K.A., Lorentzen, R.J., Kovach, R.M., & Bucci, T.J. (2001) *Environ. Health Perspect.* **109**, 277

(32)   Carlson, D.B., Williams, D.E., Spitsbergen, J.M., Ross, P.F., Bacon, C.W., Meredith, F.I., & Riley, R.T. (2001) *Toxicol. Appl. Pharmacol.* **172**, 29

(33)   Qui, M., Liu, X., & Wang, Y. (2001) *Food Addit. Contam.* **18**, 263

(34)   Ribar, S., Mesaric, M., & Bauman, M. (2001) *J. Chromatogr.* **754**, 511

(35)   Van der Westhuizen, L., Brown, N.L., Marasas, W.F.O., Swanevelder, S., & Shephard, G.S. (1999) *Food Chem. Toxicol.* **37**, 1153

(36)   Sewram, V., Nair, J.J., Nieuwoudt, T.W., Gelderblom, W.C., Marasas, W.F., & Shephard, G.S. (2001) *J. Anal. Toxicol.* **25**, 450

(37)   Watson, S.A., Mirocha, C.J., & Hayes, A.W. (1984) *Fundam. Appl. Toxicol.* **4**, 700.

(38)   Pier, A.C. (1981) *Adv. Vet. Sci. Comp. Med.* **25**, 185

(39)   Kristensen, P., Andersen, A., & Irgens, L.M. (2000) *Scand. J. Work Environ. Health* **26**, 331

(40)   Friedman, L. (1990) in *Biodeterioration Research 3.* G.C. Llewellyn & C.E. O'Rear (Eds), Plenum Press, New York, NY, 21

(41)   Porter, J.K. (1995) *J. Anim. Sci.* **73**, 871

(42)   Bacon, C.W., Lyons, P.C., Porter, J.K., & Robbins, J.D. (1986) *Agron. J.* **78**, 106

(43)   Burfening, P.J. (1973) *J. Am. Vet. Med. Assoc.* **163**, 1288

(44)   Van Rensburg, S.J., & Altenkirk, B. (1974) in *Mycotoxins*. I.F.H. Purchase (Ed), Elsevier, New York, NY, 69

(45)   Derneke, T., Kidane, Y., & Weihib, B. (1979) *Ethiop. Med. J.* **17**, 107

(46)   Krishnamachari, K.A.V.R., & Bhat, R.V. (1976) *Indian J. Med. Res.* **64**, 1624

(47)   Beardall, J.M., & Miller, J.D. (1994) in *Mycotoxins in Grains: Compounds Other Than Aflatoxin,* J.D. Miller & H.L. Trenholm (Eds), Eagan Press, St. Paul, MN, 487

(48)   *Physicians Desk Reference* (1997) 51st Ed., Medical Economics Co., Montvale, NJ, 1543

(49)   Berde, B., & Sturmer, E. (1978) in *Handbook of Experimental Pharmacology: Ergot Alkaloids and Related Compounds. Vol. 49,* B. Berde & H.O. Schild (Eds), Springer-Verlag, New York, NY, 1

(50)   Stoloff, L. (1976) in *Mycotoxins and Other Fungal Related Problems*, J.V. Rodricks (Ed), Adv. Chem. Ser. 149, American Chemical Society, Washington, DC

(51)   Etzel, R.A., Montana, E., Sorenson, W.G., Kullman, G.J., Allan, T.M., Dearborn D.G., Olson, D.R., Jarvis, B.B. & Miller, J.D. (1998) *Arch. Pediatr. Adolesc. Med.* **152**, 757

(52)   Trout, D., Bernstein, J., Martinez, K., Biagini, R., & Wallingford, K. (2001) *Environ. Health Perspect.* **109**, 641

(53)   Robbins, C.A., Swenson, L.J., Nealley, M.L., Gots, R.E., & Kelman, B.J. (2000) *Appl. Occup. Environ. Hyg.* **15**, 773

(54)   Le Bars, J., & Le Bars, P. (1990) *Assoc. Fr. Techniciens Alimentation Animale* **90**, 40

(55)   Suetens, C., Moreno-Reyes, R., Chasseur, C., Mathieu, F., Begaux, F., Haubruge, E., Durand, M.C., Nève, J., & Vanderpas, J. (2001) *Int. Orthop.* **25**, 180

(56)   Riley, R.T. (1998) in *Mycotoxins in Agriculture and Food Safety*, K.K. Sinha & D. Bhatnagar (Eds), Marcel Dekker, New York, NY, 227

(57)   Turner, W.B. (1978) *Fungal Metabolites*, Academic Press, London, England

(58)   Turner, W.B., & Alderidge, D.C. (1983) *Fungal Metabolites II*, Academic Press, London, England

(59)   Esser, K., & Lemke, P.A. (1996) in *The Mycota Vol. VI*, K. Esser & P.A. Lemke (Eds), Springer-Verlag, Berlin, Germany

4. DIETARY MYCOTOXICOSES IN NIGERIA

Tinuade A. Ogunlesi[1], Momodu-Segiru Momodu[2], John N. Ikeorah[3] and Oyuku A. Oyelami[4*]

[1]Department of Pediatrics
College of Health Sciences
Olabisi Onabanjo University
Sagamu, Nigeria.

[2]National Agency for Food and Drug Administration and Control (NAFDAC),
Federal Secretariat, Phase II, 2nd Floor, Room 219
Ikoyi, Lagos, Nigeria
Lagos, Nigeria.

[3]Nigerian Stored Products Research Institute,
Federal Ministry of Agriculture and Rural Development,
PMB 5044, Ibadan, Oyo State, Nigeria

[4*]Corresponding author
Department of Paediatrics,
Obafemi Awolowo University Teaching Hospitals Complex,
Ile Ife Wesley Guild Hospital Unit, Ilesa
PMB 5538, Ile-Ife
Osun State, Nigeria.

Since the pioneering description of the health implications of exposure to mycotoxins in the form of Turkey-X disease which followed the consumption of feeds contaminated with aflatoxins by poults on an English farm in 1960 (1), a lot of research has been conducted to further define the extent of the problem of aflatoxicosis as well as to determine the health and economic implications of contamination of foods and feeds by mycotoxin-producing fungi. Corn, processed in different forms, is a common staple food in Nigeria. Aflatoxins are ubiquitous in tropical regions and have been found in both raw and cooked and ready to eat forms of corn (2-4). They have also been found in virtually most other locally grown crops, especially cereals, spices, and nuts which, incidentally, form the bulk of staple foods in Nigeria.

With the poor economic condition of Nigeria, less than ideal conditions of production and storage of agricultural commodities are common allowing fungal and mycotoxic contamination of food crops. Storage is inevitable since reserves must be kept to last through the non-harvesting periods. Unfortunately, storage has been reported to increase the rate of fungal contamination as well as aflatoxin concentration in both raw and processed food (5-8). These poor quality foods are what most families, especially those in the low socioeconomic class, are likely to consume since they are usually offered for sale at cheap prices. A compounding predisposing factor to high scale consumption of contaminated foods and products is ignorance. Most people do not seem to pay any special attention to the obvious moldy growth on what they consume. The moldy food is not locally known to be harmful, hence, the fungi and their toxic metabolites are innocently ingested without any inkling that such may be harmful (9). Fungal contamination also occurs in stored herbal plants and preparations. These contaminated herbal preparations stored over many years are used for their medicinal values (10,11).

When ingested, mycotoxins are deposited in most tissues of the body including the liver, kidneys, brain, and the immune system and endocrine systems (12-14). This deposition has been suggested to be the rate-limiting process in the establishment of diseases of those tissues and organs. With the high prevalence of dietary mycotoxins, it is not surprising that they are also found in body fluids including sera, bile, breast milk and urine (15,16). The effects of chronic ingestion of mycotoxins manifest in the form of severe diseases affecting all ages to different extents. Mycotoxicosis refers to the effects and manifestations of acute or chronic ingestion of mycotoxin-contaminated foods and animal products. The aflatoxins are the most extensively studied mycotoxins, especially their effects on nutrition as well as hepatic and reproductive health.

MYCOTOXINS AND CHILD HEALTH IN NIGERIA

Studies in Nigeria have shown that aflatoxins are deposited in the body tissues of children. They were found in significant concentrations in the liver, kidneys, lungs and brain of dying children (12-14). The exposure of the average Nigerian child to aflatoxins begins *in-utero* as a consequence of maternal consumption of the toxins in the diet. Studies in Sierra Leone, another West African country, have shown that 58% and 20% of children had aflatoxins and ochratoxin A respectively, in their cord blood (17). Further evidence of early exposure of children to these toxins was found in a study conducted in Ibadan, Nigeria, where cord blood of infants was found to contain aflatoxins. This finding was related to unexplained cases of neonatal jaundice in the same center where the blood of 27% of newborn infants with neonatal jaundice reportedly contained aflatoxins (18,19). Similar relationships between fetal exposure to aflatoxins and neonatal jaundice were also reported by Ahmed and coworkers in Zaria, Northern Nigeria (20,21).

In addition to *in-utero* exposure, some Nigerian children are delivered with aflatoxins acquired transplacentally from the mother and they are fed breastmilk that is potentially contaminated by mycotoxins. In different Sierra Leonean study, 91% of breast milk samples were found to contain various forms of mycotoxins in concentrations greater than the levels permissible even for animal feeds (16). Most regrettably, children in most rural Nigerian communities are weaned onto diets that are based on cereals especially, corn and millet, which may be contaminated by mycotoxins.

Kwashiorkor

Acceptance of the possibility that aflatoxins could be responsible for some diseases in children followed the observation made by Hendrickse (22) that there were similarities in the geographic and climatic prevalence of kwashiorkor and exposure to dietary aflatoxins. Although the etiology of kwashiorkor is suspected to be multi-factorial, the role of aflatoxins has been extensively investigated. Kwashiorkor has been observed to occur in areas where food handling and storage facilities are poor. Furthermore, the peak of kwashiorkor occurs between the months of July and August which coincides with the peak of the rains when foods are abundant (22). This season is also ordinarily characterized by climatic features that promote the abundant growth of fungi. In fact, the Yoruba of western Nigeria called kwashiorkor, *Ile tutu* i.e. a disease that is prevalent during the damp, wet season.

It is also postulated that the high prevalence of diarrheal diseases with dehydration at this time consequent upon the consumption of contaminated surface water may allow the accumulated toxins to overwhelm the hepatic conjugating system (22). Oyelami (23) described one of a set of twins that was exclusively breastfed but died from kwashiorkor following diarrhea. Aflatoxins were found in tissues at post-mortem examination raising the possibility of aflatoxicosis being a contributor to the pathogenesis of kwashiorkor among others. This raises questions about the hypoproteinaemia theory since the protein content of breastmilk is supposedly adequate for infants. In addition, the biochemical, metabolic and immunologic derangements in kwashiorkor matched the features manifested by animals exposed to dietary aflatoxins (24).

A causal relationship between aflatoxin/kwashiorkor and infections is known. The presence of aflatoxins in the blood has been specifically observed to increase the number and frequency of infections in children with kwashiorkor (25). This may be explained by findings which suggest that aflatoxins, by causing extensive mitochondrial damage of lymphocytes in experimental rats, may in turn cause immunodysfunction

(26). In a Nigerian study, children who died with pneumonia were observed to have aflatoxins in their lungs (12). In The Philippines, clinical evidences that aflatoxins may have an immunsuppressive role in the context of pneumonia were also demonstrated (27). Infections and aflatoxins are two major noxa that were proposed by Golden and Ramdath (28) as provoking the accumulation of free radicals and the cascade that results in clinical kwashiorkor.

MYCOTOXINS AND HEPATIC DISEASES

Nigeria is one of the developing countries where viral hepatitis is endemic. Commercial blood donors, patients attending a Sickle Cell Disease Clinic and healthy controls drawn from the general population in Benn City, Nigeria, were all shown to have evidences of previous exposure to the Hepatitis B viral antigen (29). There is evidence that both aflatoxins and hepatitis B virus infection are potent risk factors for hepatocellular carcinoma (primary liver cell carcinoma) although the nature of their interaction may not be fully known. In a report, 50-100% of primary liver cell carcinoma was associated with Hepatitis B viral infection and the content of aflatoxins in the serum of Hepatitis BsAg positive individuals was higher than in seronegative cases (30). The Hepatitis B surface antigen titer and serum aflatoxins levels were found to be particularly higher among adults with primary liver cell carcinoma than controls in Ibadan.

Comparison of the general population within Ibadan, a capital city, and a neighboring rural community in southwestern Nigeria showed similar high serum aflatoxin and Hepatitis BsAg titer in both communities (31,32). Similar findings were also recorded in other studies conducted in other parts of Nigeria as well as outside Nigeria (33-35). Therefore, the carcinogenicity of aflatoxins may explain the high prevalence of primary liver carcinoma in places where Hepatitis B virus is endemic and ingestion of aflatoxins is high. Previous hepatic aflatoxins deposition may attract the Hepatitis B viral incursion which would result in hepatitis and progression to hepatocellular carcinoma. It is also possible that pre-existing hepatitis especially that due to hepatitis B virus, progresses to hepatocellular carcinoma when aflatoxin deposition on the diseased organ takes place. What is clear, however, is that, the high prevalence of hepatocellular carcinoma in some parts of Nigeria may be related to the high endemicity of Hepatitis B virus and the high serum level of aflatoxins.

Attempts have been made to explain the causal relationship between aflatoxins and hepatocellular carcinoma at the molecular level. This may include the activation of cellular oncogenes, inactivation of tumor suppressor genes, over expression of certain growth factors and specific mutations of p53 tumor suppressor gene (36). The latter possibility was described locally by Ndububa and coworkers (33).

OTHER EFFECTS OF MYCOTOXINS

Exposure to aflatoxins has also been associated with male infertility (37). On the other hand, it is likely that aflatoxins may, in certain circumstances, be useful. Hendrickse (24) postulated that exposure to aflatoxins might have a protective role against severe malaria since the illness is a very rare finding in children with kwashiorkor. The exact mechanism is unknown but it may be that the toxins create a milieu that is inimical to the proliferation of the malarial parasite. This hypothesis should be of further research interest.

PREVENTION AND CONTROL OF EXPOSURE TO MYCOTOXINS

Educating people involved in the agricultural sector, in practices such as seed fumigation, early harvesting, rapid drying and better storage systems is an important way of preventing mycotoxin contamination of food and feeds. Processing may also reduce the aflatoxin content of contaminated foodstuffs. Microwave heating has been reported to destroy pure and foodborne aflatoxin but this technique may only be useful in an industrial setting. This method is not likely to be within the reach of the people who are likely to be heavy consumers of mycotoxins (37). A more applicable process is perhaps fermentation which, apart from improving the taste, also reduces the aflatoxin content of foods and feeds (38). The potential of L- ascorbic acid to reduce aflatoxin–induced toxicity by converting the potent AFB_1 and AFG_1 to the less potent forms AFB_2 and AFG_2 may be very useful since natural sources of ascorbic acid abound especially in the tropics (38). The use of adsorbents may also be explored. Calcium montmorillonite clay (HSCAS) is an enterosorbent that has been tested in animals as a dietary supplement. It rapidly binds aflatoxins in the gastrointestinal tract thus reducing their bioavailability (36). It may find usefulness in humans as well.

Mycotoxins Monitoring

Only recently has Nigeria focused on routine regulatory control of mycotoxins to promote consumer safety and the quality of export commodities. This effort is being spearheaded by the National Agency for Food and Drug Administration and Control (NAFDAC) and other relevant government institutions. In collaboration with the Standards Organization of Nigeria (SON), various mycotoxins regulations are under review. A set of some of the regulations has been submitted to the appropriate authorities for endorsement and should be gazetted soon. This will empower the enforcement agency and increase compliance.

In collaboration with the International Atomic Energy Agency of the United Nations (IAEA), a new initiative on regulatory control and monitoring of mycotoxins was started in 1999. This has resulted in personnel training and the setting up of a laboratory capable of analyzing the major mycotoxins in Nigeria, particularly the aflatoxins and ochratoxins.

The Nigerian Stored Products Research Institute (NSPRI) is actively collaborating with NAFDAC, the private sector and some universities to promote mycotoxin programs. Aflatoxin analysis of export grade cashewnuts, vegetable oils, and peanuts is routinely conducted by NSPRI. In addition, a number of scientists and students have undergone mycotoxin analysis training under the auspices of NSPRI and an operational laboratory has been set-up at one of the collaborating universities.

Other NSPRI activities include technical assistance to students to carry out various mycotoxin related research projects including:-

(a) A market survey for aflatoxin in corn and peanuts consumed locally in the southwestern States of Nigeria.

(b) Effect of fermentation on aflatoxin content of contaminated corn.

(c) Effect of roasting on aflatoxin content of contaminated peanuts.

(d) Detoxification of contaminated food products using ether extracts of the spice *Afflamummon danielli.*

(e) A survey on the current level of aflatoxin in locally produced and utilized peanut cake.

Also planned is a regional international workshop on contaminants and residues including mycotoxins in food and feeds in line with the U.S. FDA International Workshop on Mycotoxins worldwide goals to minimize the impact of mycotoxins on health and trade.

From a health perspective, Nigerian researchers are focusing on dietary mycotoxicoses. Efforts are underway to establish any relationships between malnutrition, malaria and mycotoxins. A risk assessment procedure is also being put in place by NAFDAC. Food for local consumption and export will meet basic safety criteria and increase economic earnings when effective regulation is in place in Nigeria.

REFERENCES

(1)   Peraica M, Radic B, Luici A and Pavloic M. (1999) *Bull. World Health Organ.* **1**, 1

(2)   Oyelami O.A., Maxwell S.M., Adeoba E. (1999) *Ann Trop Paediatr.* **16,** 137

(3)   Jespersen L, Halm M, Kpodo K, Jakobsen M. (1994) *Int J Food Microbiol.* **24,** 239

(4)   Adebajo L.O., Idowu A.A., Adesanya O.O. (1994) *Mycopathologia* **126,** 183

(5)   Bankole S.A., Eseigbe D.A., Enikuomehin O.A. (1995) *Mycopathologia* **132,** 155

(6)   Adebajo L.O., Bamgbelu O.A., Olowu R.A. (1994) *Nahrung* **38,** 33

(7)   Adebajo L.O., Idowu A.A. (1994) *Mycopathologia* **126,** 21

(8)   Adebajo L.O. (1993) *Mycopathologia* **124,** 41

(9)   Bankole S.A. and Adebanjo A.S. (2003) *Afr. J. Biotechnol.* **2,** 254

(10)   Efuntoye M.O. (1999) *Mycopathologia* **147,** 43

(11)   Oyebola O.O. (1983). *Afr. J. Med. Med. Sci.* **12,** 57

(12)   Oyelami O.A., Maxwell S.M., Adelusola K.A., Aladekomo T.A., Oyelese A.O. (1997) *J. Toxicol. Environ. Health* **51,** 623

(13)   Oyelami O.A., Maxwell S.M., Adelusola K.A., Aladekomo T.A., Oyelese A.O. (1998) *J. Toxicol. Environ. Health* **55,** 31

(14)   Oyelami O.A., Maxwell S.M., Adelusola K.A., Aladekomo T.A., Oyelese A.O. (1995) *Mycopathologia* **132,** 35

(15)   Onyemelukwe G.C., Ogbadu G. (1981) *Trans. R. Soc. Trop. Med. Hyg.* **75,** 780

(16)   Jonsyn F.E., Maxwell S.M., Hendrickse R.G. (1995) *Mycopathologia* **131,** 121

(17)   Jonsyn F.E., Maxwell S.M., Hendrickse R.G. (1995) *Ann. Trop. Paediatr.* **15,** 3

(18) Maxwell S.M., Familusi J.B., Sodeinde O., Chan M.C., Hendrickse R.G. (1994) *Ann. Trop. Paediatr.* **14,** 3

(19) Sodeinde O., Chan M.C., Maxwell S.M., Familusi J.B., Hendrickse R.G. (1995) *Ann. Trop. Paediatr.* **15,** 107

(20) Ahmed H., Hendrickse R.G., Maxwell S.M., Yakubu A.M. (1995) *Ann. Trop. Paediatr.* **15,** 11

(21) Ahmed A., Hendrickse R.G., Yakubu A.M. and Maxwell S.M. (1995) *Niger. J. Paediatr.* **22,** 3

(22) Hendrickse R.G. (1985) *Kwashiorkor : 50 Years of Myth and Mystery. Do Aflatoxins Provide a Clue ?* Second PH Van Thial Lecture of the Institute of Tropical; Medicine, Rotterdam, Leiden. Foris Publications, Dordrecht, 1985.

(24) Hendrickse R.G. (1997) *Ann. Trop. Med. Parasitol.* **91,** 787

(23) Oyelami O.A., Maxwell S.M., Aladekomo T.A., Adelusola K.A. (1995) *Ann. Trop. Paediatr.* **15,** 217

(25) Adhikari M., Gita Ramjee Berjark P. (1994) *Nat. Toxins* **2,** 1

(26) Rainbow L., Maxwell S.M., Hendrickse R.G. (1994) *Mycopathologia* **125,** 33

(27) Denning D.W., Quiepo S.C., Altman D.G., Makarananda K, Neal G.E., Camallere E.L., Morgan M.R. and Tupasi T.E. (1995) *Ann. Trop. Paediatr.* **15,** 209

(28) Golden M.H.N., Ramdath D.D., Golden B.E. (1991) in *Trace Elements, Micronutrients and Free Radicals,* Dreosi I.E. (Ed), Humana Press, New Jersey, 199

(29) Mutimer D.J., Olomu A., Skidmore S., Olomu N., Ratcliffe D., Rodgers B., Mutimer H.P., Gunmson K. and Elias F. (1994) *QJM* **87,** 407

(30)   Henry S.H., Bosch F.X., Bowen J.C. (2002) *Adv. Exp. Med. Biol.* **504,** 229

(31)   Olubuyide I.O., Maxwell S.M., Hood H., Neal G.E., Hendrickse R.G. (1993) *Afr. J. Med.Med. Sci.* **22,** 89

(32)   Olubuyide I.O., Maxwell S.M., Akinyinka O.O., Hart C.A., Neal C.E., Hendrickse R.G. (1993) *Afr. J. Med.Med. Sci.* **22,** 77

(33)   Ndububa D.A., Yakicier C.M., Ojo O.S., Adeodu O.O., Rotimi O., Ogunbiyi O. and Ozturk M. (2001) *Afr. J. Med. Med. Sci.* **30,** 125

(34)   Ojo O.S., Ako-Nai A.K., Thursz M., Ndububa D.A., Durosinmi M.A., Adeodu O.O., Fatusi O.A., Goldin R.D. (1998) *E. Afr. Med. J.* **75,** 329

(35)   Diallo M.S., Sylla A., Sidibe K., Sylla B.S., Trepo C.R.S., Wild C.P. (1995) *Nat. Toxins* **3,** 6

(36)   Moradpour D., Blum H.E. (2000) *Zentralblatr. fur Chirurgie* **125,** 592

(37)   Ibeh I.N., Uraih N., Ogonar J.I. 1994. *Int. J. Fertil. Menop. S.* **39,** 208

(38)   Dada L.O., Muller H.G. 1983. *J. Cereal Sci.* **1,** 63

5. MYCOTOXICOSES IN INDIA: AN OVERVIEW

Bhumi N. Reddy and Chinnam R. Raghavender

Mycology and Plant Pathology Laboratory,
Department of Botany,
Osmania University,
Hyderabad - 500 007, A.P., India

Mycotoxicoses are diseases caused by mycotoxins in livestock, domestic animals and humans throughout the world. Exposure to mycotoxins is mostly by ingestion but also occurs by the dermal and inhalation routes. The susceptibility of individuals to mycotoxins varies considerably depending on species, age, sex and nutrition. Acute mycotoxicoses can cause serious and sometimes fatal diseases (1-5). The possibility of mycotoxin intoxication should be considered when a sudden acute disease occurs in a large population where there is no evidence of infection with a known etiological agent and there is no improvement in the clinical picture following treatment (6).

The toxicity of certain mushrooms has been known for a long time. However, the potential human hazard of the toxic products of other fungi was not recognized until the 1850s when the ingestion of rye infected with *Claviceps purpurea* and the clinical features of ergotism were described. This was followed by reports of other mycotoxicoses that affected man such as the identification of a syndrome associated with the ingestion of bread infected by *Fusarium graminearum,* recognition of human stachybotryotoxicosis, and studies on the association between alimentary toxic aleukia (ATA) and the ingestion of over-wintered grains infested with *Fusarium poae* and *Fusarium sporotrichioides (7).* The discovery of the hepatotoxic and hepatocarcinogenic *Aspergillus flavus* toxins, the aflatoxins, in the early 1960s quickly changed the perception of the nature of the hazard and control strategy in the whole field of mycotoxins. In spite of increasing knowledge concerning human mycotoxicoses, the majority of data available on mycotoxins and mycotoxicoses have been obtained from veterinary medicine. This paper reviews outbreaks of mycotoxicoses in India (Table 5.1) where the etiology of the disease is supported by mycotoxin analysis and identification of mycotoxin producing.

Table 5.1. Documented outbreates of mycotoxicoses in India

Mycotoxin (Disease)	Commodity	Fungus	Organism affected	Symptoms	Location	Year	Ref.
Ergot alkaloid (Ergotism)	Pearl millet	*Claviceps fustiformis*	Humans	Nausea, vomiting, giddiness	Western India (southern Bombay state)	1958, 1975, 1976	(13) (9) (12)
Aflatoxin B_1, B_2 (Aflatoxicosis)	Corn, peanuts	*Aspergillus flavus, A. parasiticus*	Humans Dogs Children	Rapidly developing ascites, edema of the lower limgs, portal hypertension, higher mortality rate, Cirrhosis in children	Banswada and Panchmahals districts of Rajasthan and Gujarat respectively	1974, 1975	(21) (71), (18) (19) (74)
			Chickens	Hepatomeagaly	South Canara district of Karnataka	1974, 1976	(22)
	peanut cake			High mortality	Ranga Reddy district, Andhra Pradesh Chittoor district, Andhra Pradesh Warangal, Andhra Pradesh Mysore, Karnataka	1994 1982 1985	(30) (31) (32) (33)
			Fowls	Fatty liver syndrome			
			Murrah buffaloes	Mortality in younger chicks	Andhra Pradesh	1966	(34)
			Humans, ruminants	Post-mortem lesions and liver damage	Kakinada, Andhra Pradesh, Mysore, Karnataka	1965	(29)
Aflatoxin M_1 (Aflatoxicosis)	Buffalo milk			Liver damage, subcutaneous hemorrhage		1973 1966 1969	(35) (36) (37)
Fumonisin (Fumonisin Toxicosis)	Maize and sorghum	*Fusarium moniliforme*	Humans	Transient abdominal pain, bordorygmus and diarrhea	Deccan plateau in south India	1995	(62)
			Hens	Diarrhea, reduction in food intake, egg yield, weight loss	Andhra Pradlesh	1996	(63)
T-2 toxin (ATA)	Bread (wheat)	*Fusarium sporotrichioides*	Humans	Abdominal pain, inflammation of throat, diarrhea, bloody stools, vomiting	Kashmir	1987	(64)
DON (Emetic syndrome)	Bread (wheat)	*Fusarium graminearum, F. culmorum*	Humans	Vomiting, gastrointestinal disorders	Kashmir valley	1988	(64)
Rhizopus toxin	Pearl millet	*Rhizopus nigricans*	Humans	Polyuria, thirst, anorexia, weakness and fatigue	Maharashtra	1969	(67)
Ochratoxin A	Tumeric, ginger, coriander, pepper	*Fusarium moniliforme*	Humans	Renal toxicity, nephropathy, immunosupression	Andhra Pradesh	2000	(71)

ERGOTISM

One of the earliest reported outbreaks of mycotoxicoses in the world is ergotism. Several outbreaks of ergotism in man with convulsive and gangrenous symptoms have been described in Europe from the 9th to 14th century due to consumption of bread made of rye contaminated with ergot toxin. The first symptom was a prickly sensation in the limbs, followed by swelling due to intense heat and cold. Peripheral vascoconstriction resulted in gangrene and limb loss. In the middle ages, this was known as St. Anthony's fire because it was often cured by a visit to the shrine of St. Anthony, which happened to be in an ergot free region of France. A convulsive form of ergotism involving the nervous system occurred in Europe from the late 16th to the late 19th century.

Ergot sclerotia contain a number of alkaloids, some of which are pharmaceutically important. Ergot alkaloids of rye have been thoroughly investigated and are generally classified into three main groups, i.e. ergotamine, ergometrine and ergotoxine (Table 5.2) (8). After a comparative study of the alkaloids of ergot of rye, wheat and pearl millet, Bhat and coworkers (9, 10) concluded that alkaloids in pearl millet ergot mainly belong to the clavine group in contrast to ergotamine and ergotoxine groups of alkaloids in ergots of rye and wheat. The clavine alkaloid content of ergot of pearl millet was found to vary in different agroclimatic regions (11).

Table 5.2. Molecular formulae and melting points of selected ergot alkaloids (ergolines)[a]

Group	Alkaloid	Formula	Melting point[b] °C
Ergometrine [c]	Ergometrine	$C_{19}H_{22}O_2N_3$	162
Ergotamine	Ergotamine	$C_{33}H_{35}O_5N_5$	180
	Ergosine	$C_{30}H_{37}O_5N_5$	220-230
	Ergocristine	$C_{35}H_{39}O_5N_5$	160-175
Ergotoxine	Ergocryptine	$C_{32}H_{41}O_5N_5$	212-214
	Ergocornine	$C_{31}H_{39}O_5N_5$	182-184
Ergoclavine	Agroclavine	$C_{16}H_{18}N_2$	206
	Elymoclavine	$C_{16}H_{18}ON_2$	249

Table adapted from EFSA (76)
[a]From: van Rensberg and Altenkirk (75) and Lorenz (15).
[b]Most of the ergot alkaloids decompose at melting temperatures.
[c]Also denoted ergonovine, ergobasine and ergotoxine.

Ergot contamination of pearl millet has become more widespread globally following the introduction of less ergot-resistant high yielding varieties. Ergot contamination of pearl millet by *Claviceps fusiformis* cause a type of mycotoxicosis called enteroergotism. In recent years, cases of enteroergotism manifesting as gastrointestinal disturbances from ergoty pearl millet have occurred in parts of western India (12, 13). Ergotism has been occasionally reported in the rural areas of Maharashtra, Gujarat and Rajasthan, especially in areas where high yielding varieties of pearl millet are being cultivated. An investigation of the outbreaks of ergotism in man in Sikar and Jaipur districts of Rajasthan revealed that clinical manifestations were present when the level of contamination of ergot grains in pearl millet exceeded 1.5% w/w. Krishnamachari and Bhat (12) noted that the levels of clavine alkaloids responsible for causing the disease in man appeared to be more than 26 mg/kg of pearl millet. Symptoms were usually observed within a few hours after consumption of the ergoty pearl millet and a single meal was enough to cause the disease. The clinical picture following pearl millet consumption was characterized by nausea, repeated vomiting, giddiness followed by drowsiness and a prolonged phase of sleepiness that extended over 24-48 hours. The clinical manifestations of ergot toxicity in India are different from those of classical ergotism described from Europe. The European ergotism from rye and wheat is characterized by convulsions and gangrene. Experimental ergotism in monkeys has been produced by intraperitoneally injecting ergot of pearl millet alkaloids, although the clinical manifestations were somewhat different from those in man (10).

The people in the affected areas appear to have become aware of the cause and generally resort to removal of the contaminated sclerotia (12). Many villagers have also successfully utilized the salt floatation technique that involves washing ergotized pearl millet in 5-10% common salt solution. The occurrence of the disease could further be prevented by the cultivation of ergot-resistant varieties of pearl millet or through suitable agronomic practices such as changing the sowing season, winnowing, sieving and handpicking.

On a global scale, ergotism is extremely rare today and large-scale epidemics of human poisonings due to consumption of bread prepared from ergot-contaminated grain no longer occur primarily because the normal grain cleaning and milling processes remove most of the ergot alkaloids that might otherwise end up in flour (14) or only low levels of alkaloids remain in the resultant flours. In addition, the alkaloids that enter the flour are usually destroyed during baking and cooking because of their thermo-sensitive nature (6). Furthermore, strict grading standards applied in many countries helped not to permit contaminating grain to reach commercial food channels (15). Nevertheless localized human ergotism may still happen due to ignorance or negligence.

HUMAN AFLATOXICOSES

Aflatoxicosis is primarily a hepatic disease in humans. Consumed chronically aflatoxins may cause liver cancer. Aflatoxins B_1, M_1 and G_1 have been shown to cause various types of cancers in different animal species. Epidemiological studies have demonstrated a cause effect-relation between aflatoxins but only aflatoxin B_1 is considered by the International Agency for Research on Cancer (IARC) as having produced sufficient evidence of carcinogenicity in experimental animals (16). Although humans and animals are susceptible to the effects of acute aflatoxicosis, the chances of human exposure to acute levels of aflatoxin is remote in well-developed countries. In undeveloped countries, human susceptibility can vary with age, health and level and duration of exposure.

In 1974, an outbreak of hepatitis due to aflatoxicosis was reported in 200 villages in western India (Banswada and Panchamahals districts of Rajasthan and Gujarat respectively) resulting in 106 deaths. The outbreak lasted 2 months and was confined to a tribal population whose staple food was corn. Analysis of maize samples from the region showed that affected people might have consumed food contaminated with aflatoxins at between 2,000-6,000 µg/kg daily over a period of 1 month (17, 18). Symptoms of the disease included rapidly developing ascites, edema of the lower limbs, portal hypertensions and a high mortality rate. An independent investigation (19) of the same outbreak confirmed that aflatoxins were the major cause of this disease. Toxic hepatitis was reported in three adjoining districts of north-west India during the period November and December, 1974 affecting both humans and dogs. Hepatic histology was similar to that described by Krishnamachari and co-workers (17, 18) in addition to a marked high fever at the onset of disease. The level of aflatoxin contamination in samples from affected and unaffected households was only 0.1μg/kg. It was thought that other mycotoxins or an infection might have caused the outbreak. A follow-up study of the Banswada and Panchamahals districts outbreak conducted after a period of one year (20) found that survivors were fully recovered with no ill effects from the poisoning. Childhood cirrhosis, a clinical condition mainly confined to the Indian sub-continent has been attributed to aflatoxin contamination (21). In an investigation of a third outbreak (22, 23), a correlation between aflatoxin contamination and hepatomegaly in children was reported in south Canara district of Karnataka.

In other regions of the world acute aflatoxicosis outbreaks have also occurred. Recently in eastern Kenya (24), an aflatoxicosis outbreak implicating aflatoxin-contaminated corn resulted in 317 cases and 125 deaths. Although aflatoxicosis outbreaks have occurred periodically in Africa and Asia, this outbreak resulted in the largest number of fatalities ever documented.

Although human exposure to aflatoxin (aflatoxin M_1) occurs primarily through the milk and milk products from animals that have consumed contaminated feed, aflatoxins have been shown to cross the placental barrier thus exposing the fetus to aflatoxin. Denning *et al.*, (25) reported finding aflatoxin B_1 and G_1 in human cord sera obtained at birth and in serum obtained immediately after birth from the mother. The subjects of the study were residents of Songkhla, Thailand. Of the 35 samples of cord sera, 17 (48%) contained aflatoxin in concentrations from 0.064 to 13.6, mean 3.1 nmol/ml. By comparison only two (6%) of 35 maternal sera contained aflatoxin (mean 0.62 nmol/ml). These results demonstrate transplacental transfer and concentration of aflatoxin by the feto-placental unit, which may be of biological importance. Furthermore, sera from subjects in the United Kingdom, Nigeria and Nepal were studied (26). No aflatoxin was found in U.K. sera, whilst 76% and 100% respectively of Nigerian and Nepalese samples were found positive for aflatoxin. Maternal and cord sera from Thai subjects showed that 6% of maternal blood had detectable aflatoxin whilst 49% of cord sera samples were positive for aflatoxin. This is evidence of trans-placental transfer of aflatoxin in humans and possibly of concentration of aflatoxin by the feto-placental unit. It has been observed that only a small percentage (0.092% – 0.43%) of dietary aflatoxin intake was excreted in milk (27). The IARC concluded in 1993 that there was sufficient evidence in experimental animals for the carcinogenicity of M_1 and inadequate evidence in the case of humans. No additional toxicological information on aflatoxin M_1 has appeared in the literature after 1993 (4).

Besides endangering human health, aflatoxin contamination seriously affects the export potential of high value commodity crops, such as edible nuts (peanuts, pistachios, cashewnuts and almonds) and spices (turmeric and chilies), which could provide an important source of income for farmers in the regions of semi-arid tropics. Preventing aflatoxin contamination will enable subsistence farmers to benefit from increased trade. It will also contribute to an improvement in the general health of people; often the poor who consume contaminated foods. Aflatoxins act very slowly and prolonged consumption can lead to liver cancer in humans. A person's chances of developing cancer are compounded significantly if he/she also carries the hepatitis B virus, which causes jaundice. In India, an estimated 20 million people are hepatitis B carriers. More importantly, the most commonly used food products are frequently contaminated by aflatoxin. One of the major drawbacks in enhancing aflatoxin prevention programs in India is the fact that there is no awareness about these toxins and a lack of strict enforcement of regulations. Western countries, on the other hand, have strict regulations governing the testing of food products for aflatoxins (28).

ANIMAL AFLATOXICOSIS AND FEED TO TISSUE TRANSMISSION

There are several recorded incidences of animal mycotoxicoses in India. A suspected outbreak of peanut poisoning affecting 24 *Murrah* buffaloes and resulting in the deaths of six buffaloes was reported by Sastry *et al.,* (29). This was the first record of peanut toxicity in India although frequent cases of liver damage in Murrah buffaloes without an identified causative agent had been recorded in certain areas. More than 200,000 broiler chickens died in 1994 in Ranga Reddy district of Andhra Pradesh, after eating aflatoxin contaminated peanut cake-based feeds (30). Heavy mortality in chicks in Chittoor district of Andhra Pradesh due to aflatoxicosis was reported earlier (31). The peanut cake implicated in the aflatoxicosis was contaminated with aflatoxin at 3,590 μg/kg. Another outbreak of aflatoxicosis on commercial poultry farms was reported in the same district with 100% mortality (32). Aflatoxins (1,400-3,600 μg/kg) were found in samples of corn and peanut cake fed to the birds during the outbreak. Egg production dropped by 85-40% during an outbreak of aflatoxicosis in poultry in October 1985 in and around Warangal, Andhra Pradesh (33). Post-mortem examination of dead birds revealed liver lesions of varying severity. Feed samples were found to be contaminated with aflatoxin (600 μg/kg). No mortality was observed after the contaminated feed was replaced, and egg production gradually returned to normal. Occurrence of aflatoxicosis in poultry in Mysore state was also reported (34). The disease was first recognized at the Government Poultry Breeding Unit, Hebbal, Bangalore in 1966 when 2,219 chicks died in one week. Subsequently, several sporadic incidences were reported on various poultry farms in the state. The disease was predominant in younger stocks, possibly due to higher inclusion of contaminated peanut cake to afford higher protein levels in the ration.

Aflatoxin M_1 and M_2 are the hydroxylated metabolites of aflatoxins B_1 and B_2 and are produced in the milk when cows or other ruminants that ingest feed contaminated with these mycotoxins. Limited studies are available on aflatoxin M_1 occurrence in milk in India. An investigation was conducted in Andhra Pradesh by Yadagiri and Tulpule (35) on aflatoxin contamination in buffalo milk. Of 50 milk samples analyzed, 27 were contaminated with aflatoxin M_1 in amounts ranging from traces to 4.8 µg/L. Analysis of the peanut cake, which comprised 30% of the ration of cattle, indicated the presence of aflatoxin levels ranging from 1,000 to 3,000 µg/kg. Other outbreaks of aflatoxicoses were reported earlier in dairy cattle from the Mysore region (36) and in *Murrah* buffaloes from Andhra Pradesh (37). In experiments carried out with lactating monkeys, administration of 500 µg of aflatoxin per day over a period of 18 weeks did not produce any toxic effects in suckling young ones although, their mothers showed hepatic lesions attributable to aflatoxin (38).

FUMONISIN TOXICOSES

The fungus *Fusarium moniliforme* occurs worldwide on a variety of plant hosts and is the most prevalent fungi of corn (39, 40). This fungus was initially associated with the occurrence of equine leucoencephalomalacia (ELEM), a fatal disease of horses (41) as well as porcine pulmonary edema (PPE) and hydrothorax syndrome in pigs (42). A number of fumonisins have been isolated and characterized (43- 45), but toxicity data has only been reported for fumonisin B_1 (FB_1). Fumonisin B_1, either in purified form or in naturally contaminated corn or corn-based feeds, has been found to be the causative agent of ELEM and PPE (46-48). Fumonisin B_1 also causes liver toxicity and liver cancer in rats, atherosclerosis in monkeys and immunosuppression in poultry (49).

Fumonisins have been found to be epidemiologically associated with a high incidence of esophageal cancer in human beings in the Transkei area of South Africa (50) and in certain provinces of China (51). Recently, fumonisin has been implicated as a possible cause of an acute disease outbreak in human beings in several areas of India (52). On the basis of available data on the toxicity and carcinogenicity of FB_1, the International Agency for Research on Cancer classified the *F. moniliforme* toxin as a group 2B carcinogen i.e., potentially carcinogenic to human beings (53).

Corn and /or corn-based feeds are the major commodities in which the natural occurrence of the FB_1 or fumonisins have been reported from many parts of the world including the USA (54), Brazil (55), Asia (56), Italy (57), Costa Rica (58) and Hungary (59).

In India, high levels of FB_1 were reported in corn kernels infected with *F. moniliforme* (60) and in corn as well as poultry feeds (61). A single outbreak of acute foodborne disease possibly caused by fumonisins reported to have occurred in 27 villages of the Deccan plateau in southern India in October 1995 affected 1,325 people (62). The outbreak was a sequel to the damage of corn and sorghum crop by two cyclonic storms, which had occurred in quick succession resulting in unseasonal rains during the harvest season. The crops, either standing or harvested and left in the field, were infected by a variety of molds, predominantly *F. moniliforme* and *Aspergillus sp.* The prevailing high moisture provided a conducive environment for mold growth and fumonisin B_1 production. When the grains were eventually processed and consumed it resulted into illness. The main features of the disease were transient abdominal pain, borborygmus and diarrhea, which began half-an-hour to one hour following consumption of unleavened bread prepared from moldy sorghum or moldy corn. Patients recovered fully when the exposure ceased and there were no fatalities. The corn and sorghum samples consumed in the affected households had higher levels of fumonisin B_1 (0.04-64.7 mg/kg and 0.07-7.8 mg/kg in corn and sorghum, respectively) than in samples from control households (63). Fumonisin has been frequently found in high concentrations in corn in regions with a higher incidence of esophageal cancer such as the Transkei in South Africa and China. Corn is a staple food in these localities.

A recent outbreak of fumonisin toxicosis in a poultry farm in Andhra Pradesh, India was reported by (63). In this outbreak, a total of 6,700 hens aged 64 weeks and 3,000 hens aged 36 weeks were affected with 10% mortality. The disease was characterized by sticky diarrhea, severe reduction in food intake, egg production and body weight followed by lameness and death. Analysis of the feed indicated contamination with fumonisin B$_1$ up to 8.5 mg/kg. The disease was reproduced in day old cockerels fed the suspect diet containing 8.5 mg/kg fumonisin and in laying hens fed a diet spiked with fumonisin B$_1$ at 8 and 16 mg/kg.

ALIMENTARY TOXIC ALEUKIA (ATA)

The trichothecenes comprise a large group of mycotoxins that are produced by a variety of *Fusarium* molds. *F. sporotrichioides,* the major producer of T-2 toxin, occurs mainly in temperate regions and is associated with cereals, which have been allowed to over-winter in the field. T-2 toxin has been implicated in two outbreaks of acute human mycotoxicoses. The first occurred in Siberia (in the former USSR) during the Second World War, producing a disease known as Alimentary Toxic Aleukia (ATA). Thousands of people who had been forced to eat grain, which had over-wintered in the field, were affected and entire villages were eliminated. The toxin involved produced consistent symptoms in consumers: abdominal pain, nausea, vomiting, chills, diarrhea, and effects at a cellular level, most notably inhibiting the division of rapidly growing cell types such as the lining of the intestine and the bone marrow. The second outbreak of trichothecene poisoning was reported in Kashmir, India, in 1987 and attributed to the consumption of bread made from moldy flour (64). T-2 toxin, deoxynivalenol (DON), nivalenol and DON monoacetate were isolated from the flour. The major symptoms were abdominal pains, inflammation of the throat, diarrhea, bloody stools and vomiting. The most significant effect of T-2 toxin and other trichothecenes may be the immunosuppressive activity, which has been clearly demonstrated in experimental animals.

EMETIC (FEED REFUSAL) SYNDROME

Deoxynivalenol (also known as vomitoxin) is a type B trichothecene that occurs predominantly in grains such as wheat, barley, oats, rye, corn and less often in rice and sorghum. The occurrence of deoxynivalenol is associated primarily with *Fusarium graminearum* (*Gibberella zeae*) and *F. culmorum*, important plant pathogens that cause *Fusarium* head blight in wheat and *Gibberella* ear rot in corn. A direct relationship between the incidence of *Fusarium* head blight and contamination of wheat with deoxynivalenol has been established (65). The incidence of *Fusarium* head blight is strongly associated with moisture at the time of flowering (anthesis), and the timing rather than the amount of rainfall. The geographical distribution of the two species appears to be related to temperature, *F. graminearum* being the commoner species and occurring in warmer climates. Deoxynivalenol has been implicated in incidents of mycotoxicoses in both humans and farm animals. Refusal to consume contaminated feedstuff is the typical sign, which limits development of other signs. If no other food is offered, animals may eat reluctantly, but in some instances, excessive salivation and vomiting may occur. In the past, the ability to cause vomiting had been ascribed to deoxynivalenol only, hence the common name, vomitoxin. However, othermembers of the trichothescene family also can induce vomiting. In swine, reduced feed intake may occur at dietary concentrations as low as 1 mg/kg, and refusal may be complete at 10 mg/kg. Ruminants generally will readily consume up to 10 mg/kg dietary vomitoxin, and poultry may tolerate as much as 100 mg/kg. Related effects of weight loss, hypoproteinemia, and weakness may follow prolonged feed refusal. There is little credible evidence that vomitoxin causes reproductive dysfunction in domestic animals.

Many acute disease outbreaks from exposure to DON have been reported in China and India (66). Deoxynivalenol was implicated in a human mycotoxicosis in India although other tricothecenes including T-2 toxin were also present in the food. From June to September, 1987, it was reported that a considerable segment of the population in the subtropical Kashmir valley was affected by a gastrointestinal disorder (64). Epidemiological investigations indicated that the outbreak was associated with the consumption of bread made from mold-damaged wheat. *Fusarium* sp. and *Aspergillus* sp. and varying quantities of trichothecene mycotoxins (DON, nivalenol, acetyldexoynivalenol, T-toxin) were in found in samples from the region. The symptoms were reproduced in dogs fed extracts from the contaminated samples. This incident demonstrated that trichothecene mycotoxins can cause an outbreak of disease in humans.

OTHER MYCOTOXICOSES

There have been reports of mycotoxicoses associated with other toxins such as *Rhizopus* toxin and Ochratoxin in India but they are of minor importance. Consumption of *Rhizopus nigricans* contaminated pearl millet is reported to cause a disease in humans characterized by polyuria, thirst, anorexia, weakness and fatigue in Maharashtra, western India (67). Similar symptoms were observed in Swiss albino rats fed a pure culture of the fungus (68). Ochratoxin A produces renal toxicity, nephropathy and immunosuppression in several animal species. A correlation between human exposure to ochratoxin A and the development of endemic nephropathy (a fatal, chronic renal disease occurring in limited areas of Bulgaria, the former Yugoslavia and Romania) has been suggested (69). Although there is currently inadequate evidence in humans for the carcinogenicity of ochratoxin A, there is sufficient evidence in experimental animals. Ochratoxin A has been found in significant quantities in pig meat, as a result of its transfer from feed (70). Thirumala Devi *et al.,* (71) reported, for the first time in India, the occurrence of Ochratoxin A in some of the most widely used spices in Indian cooking. Out of 120 samples of black pepper, coriander, ginger and turmeric, 45 samples contained ochratoxin A in the concentration range of 10-110 µg/kg. High levels of ochratoxin A were detected in turmeric, which is one of the most widely used spices in India.

SUMMARY AND CONCLUSIONS

Mycotoxins are gaining increasing importance due to their deleterious effects on human and animal health. Some mycotoxins occur more frequently in tropical countries because of high temperature, moisture, and unseasonal rains. Furthermore, diets in many developing countries are more heavily based on crops susceptible to mycotoxins. Hence mycotoxin-related chronic health risks and disease outbreaks continue to be problems of significant public health concern, particularly prevalent in developing countries like India. Ergotism is one of the earliest known outbreaks of mycotoxicosis reported in rural areas of western India associated with pearl millet. Similarly, one of the first outbreaks of aflatoxicosis was reported in western India during 1974 with 106 deaths of humans whose staple food was corn. Childhood cirrhosis mainly confined to the Indian subcontinent has been attributed to aflatoxin contamination. An outbreak of acute foodborne disease caused by fumonisins was reported in 27 villages in south India during 1995 affecting 1,375 people. This outbreak occurred after cyclones caused the infection of corn and sorghum crops and by *Fusarium moniliforme.* The consumption of bread made from moldy flour contaminated with T-2 was implicated in an acute human mycotoxicosis known as alimentary toxic aleukia (ATA) in India in 1987. Deoxynivalenol (vomitoxin) produced by *Fusarium graminearum,* was a suspected causative agent of an outbreak of emetic (feed refusal) syndrome in Kashmir. Consumption of *Rhizopus nigricans* contaminated pearl millet causing illness in humans is reported from Maharashtra.

Most of the outbreaks of the mycotoxicoses described here are consequences of the ingestion of food or feed contaminated with mycotoxins because of poor agricultural practices. Sufficient attention to such disease outbreaks has been lacking in view of the remoteness of the affected localities (19). India has a population of more than one billion and nearly 40% of the people live below the poverty line. It is very difficult to imagine that the poor of this society will have access to completely safe and toxin free food any time soon. Programs that focus on prevention of contamination will provide some relief. History can be our teacher. We learn from (71) that recognition of the association of ATA with the consumption of food contaminated by molds and corresponding preventive measures resulted in the eradication of the disease.

REFERENCES

(1)   Bennett, J.W., & Klich, M. (2003) *Clin. Microbiol. Rev.* 16, 497

(2)   Hussein, H.S., & Brasel, J.M. (2001) *Toxicology* 167, 101

(3)   IARC. (1993) *Evaluation of the Carcinogenic Risks of Chemicals to Man- Some Naturally Occurring Substances: Food Items and Constituents, Heterocyclic Aromatic Amines and Mycotoxins, IARC* Monograph 56, International Agency for Research on Cancer, Lyon, France

(4)   IARC. (2002) *Evaluation of the Carcinogenic Risks of Chemicals to Man- Some Naturally Occurring Substances: Food Items and Constituents, Heterocyclic Aromatic Amines and Mycotoxins, IARC* Monograph 82, International Agency for Research on Cancer, Lyon, France

(5)   Speijers, G.J.A., & Speijers, M.H.M. (2004) *Toxicol. Lett.* 153, 91

(6)   Peraica, M., Radic, B., Lucic, A., & Pavlovic, M. (1999) *Bull. World Health Organ.* 77, 754

(7)   Sarkisov, A.K. (1954) in *Mycotoxicoses*, Moscow State Publishing House for Agricultural Literature, Sel'khozgiz, 216.

(8)   Bove, F.J. (1970) *The Story of Ergot.* S. Karger Ag., Basel, Switzerland

(9)   Bhat, R.V., Roy, D.N., & Tulpule, P.G. (1975) *Proc. Nutr. Soc. India* 19, 7

(10)   Bhat, R.V., & Roy, D.N. (1976) *Indian J. Med. Res.* 64, 1629

(11)   Singh, H.N., & Hussain, A. (1976) *Arogya J. Health Sci.* 2, 115

(12)   Krishnamachari, K.A.V.R., & Bhat, R.V. (1976) *Indian J. Med. Res.* 64, 1624

(13)   Patel, J.B., Boman, T.J., & Dallal, V.C. (1958) *Indian J. Med. Sci.* 12, 257

(14)   Scott, P.M. & Bhat, R.V. (1997) *J. Agric. Food Chem.* 45, 2170

(15)   Lorenz, K. (1979) *CRC Cr. Rev. Food Sci. Nutri.* 11, 311

(16)   Eaton, D.L., & Groopman, J.D. (1994) in *The Toxicology of Aflatoxins,* Academic Press, New York, 383

(17)   Krishnamachari, K.A.V.R., Bhat, R.V., Nagarajan, V., & Tilak, T.B.G. (1975a) *Lancet* 1, 1061

(18)   Krishnamachari, K.A.V.R., Bhat, R.V., Nagarajan, V., & Tilak, T.B.G. (1975b) *Indian J. Med. Res.* 63, 1036

(19)   Tandon, B.N., Krishnamurthy, L., Koshy, A., Tandon, H.D., Ramalingaswami, V., Bhandari, J.R., Mathur, M.M., & Mathur, P.D. (1977) *Gastroenterology* 72, 488

(20)   Bhat, R.V., & Krishnamachari, K.A.V.R. (1977) *Indian J. Med. Res.* 66, 55

(21)   Amla, I., Kamala, C.S., Gopala Krishna, G.S., Jayaraj, A.P., Sreenivasamurthy, V., & Parpia, H.A.B. (1971) *Am. J. Clin. Nutr.* 24, 609

(22)   Sreenivasamurhty, V. (1975) *Proc. Nutr. Soc. India* 19, 1

(23)   Sreenivasamurthy, V. (1977) *Arogya J. Health Sci.,* 3, 4

(24)   Azziz-Baumgartner, E., Lindblade, K., Gieseker, K., Rogers, H.S., Kieszak, S., Njapau, H., Schleicher, R., McCoy, L.F., Misore, A., DeCock, K., Rubin, C., Slutsker, L., & The Aflatoxin Investigative Group. (2005) *Environ. Health Perspect.,* 113, 1779

(25)   Denning, D.W., Allen, R., Wilkinson, A.P., & Morgan, M.R.A. (1991) *Carcinogenesis* 11, 1033

(26)   Wilkinson, A.P., Denning, D.W., & Morgan, M.R.A. (1989) *J. Toxicol. Toxin Rev.* 8, 69

(27)   Lamplugh, S.M., Hendrickse, R.G., Ageagyei, F., & Mwanmut, D.D. (1988) *Brit. Med. J., (Clin. Res. Ed)* 296, 968

(28)   FAO. (2004). *Worldwide Regulations for Mycotoxins in Food and Feed in 2003*. FAO, Food and Nutrition Paper 81, 81

(29)   Sastry, G.A., Narayana, J.V., Rama Rao, P., Christopher, K.J., & Hill, K.R. (1965) *Indian Vet. J.* 42, 79

(30)   ICRISAT. (2002). *Aflatoxin: A deadly Hazard*, www.icrisat. org/media/2002/media9.htm.

(31)   Char, N.L., Rao, P., Khan, I., & Sarma, D.R. (1982) *Poultry Adviser* 15, 57

(32)   Choudary, C., & Rao, M.R.K.M. (1982) *Poultry Adviser* 16, 75

(33)   Choudary, C. (1986) *Poultry Adviser* 19, 59

(34)   Gopal T, Zaki, S., Narayanaswamy, M., & Premlata, S. (1969) *Indian Vet. J.,* 46, 348

(35)   Yadagiri, B., Tupule, P.G. (1974) *Indian J. Dairy Sci.,* 27, 293

(36)   Gopal, T., Syed, Z., Narayanaswamy, M., & Premlata, S. (1968) *Indian Vet. J.,* 45, 702

(37)   NIN. (1969) *Annual Report*. National Institute of Nutrition, Hyderabad, India

(38)   Mohiuddin, S.M. (1972) *Studies on the Effects of Aaflatoxin on Monkeys*. PhD Thesis, Konkan Krishi Vidyapeeth, Dapoli, India

(39)   Jindal, N., Mahipal, S.K., & Rottinghaus, G.E. (1999) *Mycopathology*. 148, 37

(40)   Marasas, W.F.O., Nelson, P.E., & Toussoun, T.A. (1984) in *Toxigenic Fusarium Species, Identity and Mycotoxicology*, Pennsylvania State University Press, University Park, PA, 216

(41)   Marasas, W.F.O., Kellerman, T.S., Pienaar, J.G., & Naude, T.W. (1976) *Onderstepoort J. Vet. Res.*, 43, 113

(42)   Kriek, N.P.J., Kellerman, T.S., & Marasas, W.F.O. (1981) *Onderstepoort J. Vet. Res.* 48, 129

(43)   Gelderblom, W.C.A., Jaskiewicz, K., Marasas, W.F.O., Thiel, P.G., Horak, R.M., Vleggaar, R., & Kriek, N.P.J. (1988) *Appl. Environ. Microbiol.* 54, 1806

(44)   Bezuidenhout, S.C., Gelderblom, W.C.A., Gorst-Allman, C.P., Horak, R.M., Marasas, W.F.O., Spiteller, G., & Vleggaar, R. (1988) *J. Chem. Soc. Chem. Commun.*, 1988, 743

(45)   Cawood, M.E., Gelderblom, W.C.A., Vleggaar, R., Behrend, Y., Thiel, P.G., & Marasas, W.F.O. (1991) *J. Agric. Food. Chem.* 39, 1958

(46)   Harrison, L.R., Colvin, B.M., Greene, J.T., Newman, L.E., & Cole, J.R. Jr. (1990) *J. Vet. Diagn. Invest.* 2, 217

(47)   Kellerman, T.S., Marasas, W.F.O., Thiel, P.G., Gelderblom, W.C.A., Cawood, M., & Coetzer, J.A.W. (1990) *Onderstepoort J. Vet. Res.* 57, 269

(48)   Marasas, W.F.O., Kellerman, T.S., Gelderblom, W.C.A., Coetzer, J.A.W., Thiel, P.G., & Vander Luzt, J.J. (1988) *Onderstepoort J. Vet. Res.* 55, 197

(49)   Norred, W.P. (1993) *J. Toxic. Environ. Health.* 38, 309

(50)   Rheeder, J.P., Marasas, W.F.O., Thiel, P.G., Sydenham, E.W., Shephard, G.S., & Van Schalkwyk, D.J. (1992) *Phytopathology* 82, 353

(51)   Yoshizawa, T., Yamashita, A., & Luo, Y. (1994) *Appl. Envrion. Microbiol.* 60, 1626

(52)   Anonymous. (1998) *ICMR Bull.* 28, 53

(53)   Vainio, H., Heseltinek, E., & Wilbourn, J. (1993) *Int. J. Cancer* 53, 535

(54)   Ross, P.F., Rice, L.G., Plattner, R.D., Osweiler, G.D., Wilson, T.M., Owens, D.L., Nelson, H.A., & Richard, J.L. (1991) *Mycopathology* 114, 129

(55)   Sydenham, E.W., Marasas, W.F.O., Shephard, G.S., Thiel, P.G., & Hirooka, E.Y. (1992) *J. Agric. Food Chem.* 40, 994

(56)   Ueno, Y., Aoyama, S., Sugiura, Y., Wang, D.S., Lee, U.S., Hirooka, E.Y., Hara, S., Karki, T., Chen, G., & Yu, S.Z. (1993) *Mycotoxin Res.* 9, 27

(57)   Ritieni, A., Moretti, A., Logrieco, A., Bottalico, A., Randazzo, G., Monti, S.M., Ferracane, R., & Fogliano, V. (1997) *J. Agric. Food Chem.* 45, 4011

(58)   Viquez, O.M., Castell-Perez, M.E., & Shelby, R.A. (1996) *J. Agric. Food Chem.* 44, 2789

(59)   Fazekas, B., Bajmocy, E., Glavits, R., Fenyvesi, A., & Tanyi, J. (1998) *J. Vet. Med.* 45, 171

(60)   Chhatterjee, D., & Mukherjee, S.K. (1994) *Lett. Appl. Microbiol.* 18, 251

(61)   Shetty, P.H., & Bhat, R.V. (1997) *J. Agric. Food Chem.* 45, 2170

(62)   Bhat, R.V., Shetty, H.P.K., Amruth, R.P., & Sudershan, R.V. (1997) *Clin. Toxicol.* 35, 249

(63)   Prathapkumar, S.H., Rao, V.S., Paramkishan, R.J., & Bhat, R.V. (1997) *Br. Poult. Sci.* 38, 475

(64)   Bhat, R.V., Beedu, S.R., Ramakrishna, Y., & Munshi, K.L. (1989) *Lancet* 1, 35

(65)   McMullen, M., Jones, R., & Gallenberg, D. (1997) *Plant Dis.* 81, 1340

(66)   Bhat, R.V., & Vasanthi. S. (2003) *Mycotoxin Food Safety Risk in Developing Countries: Food Safety, Food Security and Food Trade.* *IFPRI* Focus 10 Brief 3 of 17

(67)   Deodhar, N.S., Rao, U.N., Ganla, U.G., Sule, C.R., & Mistry, C.J. (1970) *Indian. J. Med. Sci.* 24, 626

(68)   Narasimhan, M.J. Jr., Gangla, V.G., Deodhar, N.S., & Sule, C.R. (1967) *Lancet* 1, 760

(69)   Pitt, J.I., & Leistner, L. (1991) in *Mycotoxins and Animal Foods,* Smith J.E., & Henderson, R.S., (Eds), CRC Press, Inc., 81

(70)   FAO. (1994) in *The Significance of Mycotoxins,* Proctor, D.L. (Ed), Food and Agriculture Organization, Rome, Italy

(71)   Thirumala Devi, K., Mayo, M.A., Gopal, R., Emmanuel, K.E., Larondelle, Y., & Reddy, D.V.R. (2001) *Food Addit. Contam.* 18, 830

(72)   Leonov, A.N. (1977) in *Mycotoxins in Human and Animal Health*, Rodricks, J.V., Hesseltine, C.W., & Mehlman, M.A., (Eds), Pathotox Publishers Inc., Park Forest South, IL, USA, 323

(73)   Scott, P.M., & Lawrence, G.A. (1980) *J. Agric. Food Chem.*, 28, 1258

(74)   Tandon, H.D. (1993) *Food Addit. Contam.*, 10, 105

(75)   Van Rensburg, S.J., & Altenkirk, B. (1974) in *Mycotoxins*, Purchase I.F.H., (Ed), Elsevier Science Publishers, Amsterdam, New York, Oxford: pp. 69

(76)   EFSA (2005) *The EFSA Journal* **225**, 1

Section III: Sampling and Analysis

6. MYCOTOXIN SAMPLING PLANS

Thomas B. Whitaker

U.S. Department of Agriculture, Agricultural Research Service
124 Weaver Laboratory
North Carolina State University
Raleigh, North Carolina 27695

It is important to be able to detect and quantify the mycotoxin concentration in foods and feeds destined for human and animal consumption. In research, quality assurance, and regulatory activities, correct decisions about the fate of commercial bulk lots can be made only if the mycotoxin concentration in the lot can be calculated with a high degree of accuracy and precision. The mycotoxin concentration of a bulk lot is usually estimated by measuring the mycotoxin concentration in a small portion of the lot, or a sample is taken from the lot. The mycotoxin concentration in the bulk lot is assumed to be the same as the measured mycotoxin concentration in the sample. Then, based on the measured sample concentration, a decision is made about the edible quality of the bulk lot. For example, in a regulatory environment, decisions are made to classify the lot as acceptable or unacceptable based on a comparison of the measured sample concentration with a legal limit. If the sample concentration does not accurately reflect the lot concentration, then the lot may be misclassified and there may be undesirable economic and/ or health consequences. Fortunately, sampling plans can be designed to minimize the misclassification of lots and reduce the undesirable consequences associated with regulatory decisions about the fate of bulk lots. In this chapter, sampling plans are defined, risks associated with misclassifying lots are discussed, and the designs of sampling plans that reduce risks associated with misclassification of lots are described.

DEFINITION

A mycotoxin-sampling plan is defined by a mycotoxin test procedure and a defined accept/reject limit. A mycotoxin test procedure is a multistage process that generally consists of three steps: sampling, sample preparation, and analytical steps. The sampling step specifies how the sample is selected or taken from the bulk lot as well as the size of the sample. For granular products, the sample preparation step is also a two-part process in which the sample is ground in a mill to reduce particle size and a subsample is removed from the comminuted sample. Finally, in the analytical step, the mycotoxin is solvent extracted from the comminuted subsample and quantified by approved procedures.

The measured mycotoxin concentration in the sample is used to estimate the true mycotoxin concentration in the bulk lot. In a regulatory environment, the measured mycotoxin concentration is compared with a defined accept/reject limit that usually equals a legal limit. If the measured concentration is greater than the defined accept/reject limit, the lot is rejected and diverted from the food and feed markets. Otherwise, it is accepted and is processed into food or feed products. Comparing the measured concentration with an accept/reject limit is often called acceptance sampling because the measured concentration value is not as important as whether the measured concentration (and thus the lot concentration) is above or below a legal limit. The components of a sampling plan are discussed in greater detail below.

SAMPLING

Sample Selection

Procedures used to take a sample from a bulk lot are extremely important. Every individual item in the lot should have an equal chance of being chosen (called random sampling). Biases are introduced by sample selection methods if equipment and procedures used to select the sample prohibit or reduce the chances of any item in the lot from being chosen (1). If the lot has been blended thoroughly from the various material handling operations, then the contaminated particles are probably distributed uniformly throughout the lot (2). In this situation, it is probably not too important from which location in the lot the sample is drawn. However, if the lot is contaminated because of moisture leaks that cause clumps of high moisture or for other localized reasons, then the mycotoxin-contaminated particles may be located in isolated pockets in the lot (3). If the sample is drawn from a single location, the contaminated particles may be missed or too many contaminated particles may be collected. Because contaminated particles may not be distributed uniformly throughout the lot, the sample should be an accumulation of many small portions taken from many different locations throughout the lot (4, 5). The Food and Agriculture Organization/World Health Organization (FAO/WHO) recommends that each incremental portion be about 200 g and that one incremental portion be taken for every 200 kg of product (6). The accumulation of many small incremental portions is called a bulk sample. If the bulk sample is larger than desired, it should be blended and subdivided until the desired sample size is achieved. The smallest sample size that is subdivided from the bulk sample and comminuted in a grinder in the sample preparation step is called the test sample. It is generally more difficult to obtain a representative (lack of bias) test sample from a lot at rest (static lot) than from a moving stream of the product (dynamic lot) as the lot is moved from one location to another. Sample selection methods differ depending on whether the lot is static or dynamic.

Static Lots

Examples of static lots are commodities contained in storage bins, railcars, or many small containers such as sacks. When a sample is being drawn from a bulk container, a probing pattern should be developed so the product can be collected from different locations in the lot. An example of several probing patterns used by the U.S. Department of Agriculture (USDA) to collect samples from peanut lots is shown in Figure 6.1 (7–9). The sampling probe should be long enough to reach the bottom of the container when possible. Attempts should be made to use a sampling rate similar to the 200 g per 200 kg mentioned above. However, it may not be possible to achieve the suggested sampling rate because of the design of the sampling equipment, the size of the individual containers, and the size of the lot.

1 Front

X	O	
		X
	X	O
O		X
X		

2 Front

	O	X
X		
O	X	
X		O
		X

3 Front

	X	
X	O	
		X
X		O
O	X	

4 Front

	X	
	O	X
X		
O		X
	X	O

5 Front

		X
O	X	
X		O
	X	
	O	X

6 Front

X		
	X	O
O		X
	X	
X	O	

7 Front

X		O
	X	
	O	X
X		
O	X	

8 Front

O		X
	X	
X	O	
		X
	X	O

9 Front

O		X
X		O
	X	
	O	X
X		

10 Front

X		O
O		X
	X	
X	O	
		X

11 Front

	X	
		X
X		
	X	
		X

12 Front

	X	
X		
		X
	X	
X		

x = 5 Probe Patterns
x + 0 = 8 Probe Patterns

Figure 6.1. Example of several five- and eight-probe patterns used by the USDA to sample farmers' stock peanuts for grade and support price.

When a static lot in separate containers such as sacks or retail containers is being sampled, the sample should be taken from many containers dispersed throughout the lot.

When sacks are being stored in a storage facility, access lanes should be constructed to allow access to sacks at interior locations. The recommended number of containers sampled can vary from one in four in small lots (less than 20 metric tones) to the square root of the total number of containers for large (greater than 20 metric tones) lots (6).

If the lot is in a container where access is limited, the sample should be drawn when the product is being removed from or being placed into the container. If the accumulated bulk sample is larger than required, the bulk sample should be thoroughly blended and reduced to the required test sample size with a suitable divider that randomly removes a test sample from the bulk sample.

Dynamic Lots

True random sampling can be more nearly achieved when a bulk sample is being selected from a moving stream as the product is transferred (i.e., conveyor belt) from one location to another. When samples are being taken from a moving stream, small increments of product should be taken along the entire length of the moving stream; small increments of product should be taken across the entire cross section of the moving stream; all the increments of product should be combined to obtain a bulk sample; if the bulk sample is larger than required, then it should be blended and subdivided to obtain the desired test sample size.

Automatic sampling equipment such as cross-cut samplers (Figure 6.2) are available commercially with timers that automatically pass a diverter cup through the moving stream at predetermined and uniform intervals. When automatic equipment is not available, a person can be assigned to manually pass a cup through the stream at periodic intervals to collect the bulk sample. Whether automatic or manual methods are used, small increments of product should be collected at frequent and uniform intervals throughout the time product flows past the sampling point and combined.

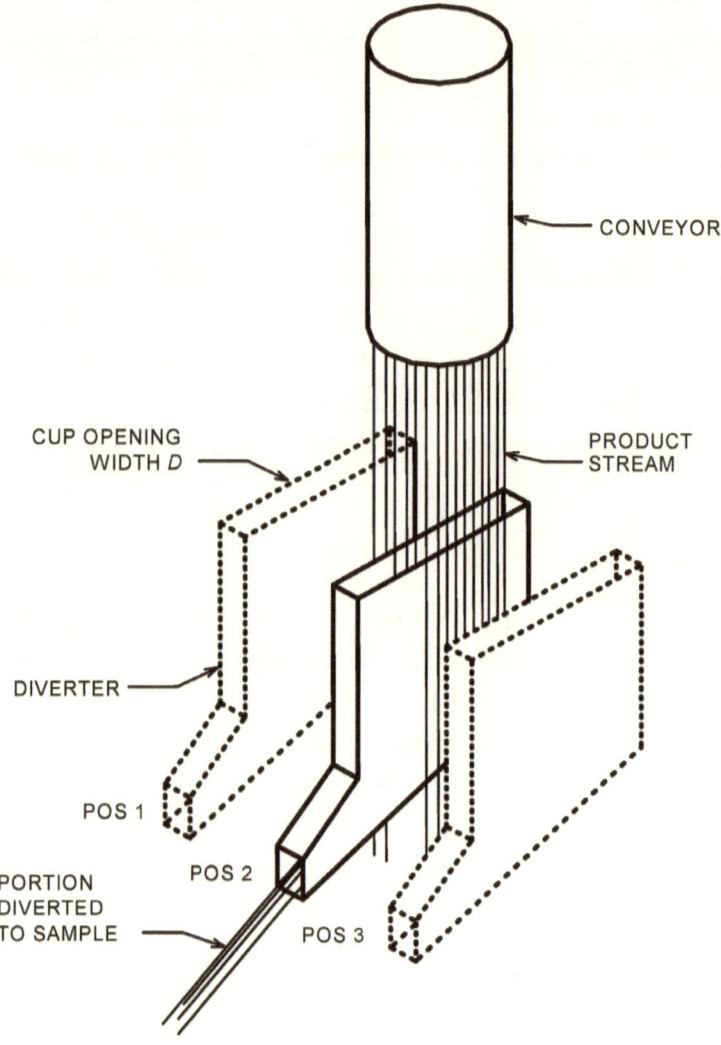

Figure 6.2. Automatic cross-cut sampler.

Cross-cut samplers should be installed in the following manner: (i) the plane of the opening of the sampling cup should be perpendicular to the direction of flow; (ii) the sampling cup should pass through the entire cross-sectional area of the stream; and (iii) the opening of the sampling cup should be wide enough to accept all items of interest in the lot. As a general rule, the width of the sampling cup opening should be two to three times the largest dimensions of the items in the lot.

The size of the bulk sample S (in kg) taken from a lot by a cross-cut sampler is

$$S = (DL)/(TV) \qquad\qquad (1)$$

where D is the width of the sampling cup opening (in cm), L is the lot size (in kg), T is the interval or time between cup movement through the stream (in sec), and V is cup velocity (in cm/sec).

Equation 1 can also be used to compute other terms of interest such as the time between cuts T; for example, the required time T between cuts of the sampling cup to obtain a 10-kg sample from a 30,000-kg lot where the sampling cup width is 5.08 cm (2 in.) and the cup velocity through the stream is 30 cm/sec. Solving for T in Equation 1,

$$T = (5.08 \text{ cm} \times 30{,}000 \text{ kg})/(10 \text{ kg} \times 30 \text{ cm/sec}) = 508 \text{ sec}$$

If the lot is moving at 1,000 kg/min, the entire lot will pass through the sampler in 30 min and only three or four cuts will be made by the cup through the lot. This may be considered too infrequent, because too much product passes the sampling point between the times the cup cuts through the stream. The interaction among the variables in Equation 1 needs to be fully understood in terms of the amount of sample accumulated and the frequency of taking the product.

Bulk Versus Test Sample

Because contaminated seed may not be uniformly dispersed throughout the lot, many incremental portions are taken from many different locations throughout the lot and accumulated to form a bulk sample. As a result, the bulk sample is usually larger than the desired test sample size used to estimate the lot mycotoxin concentration. For granular material, the test sample is the smallest sample of granular product ground in a mill in the sample preparation step. For finely ground materials (corn flour) or liquids (milk), the test sample is the smallest sample used in the analytical step to quantify the mycotoxin. When the bulk sample is larger than the test sample, dividers are used to remove the desired test sample from the bulk sample. If a test sample is to be removed from the bulk sample, then it is best to use a mechanical device such as a Boerner or riffle divider (8) that is considered to give random divisions. When a divider that gives random divisions is being used, the bulk sample does not have to be blended before the test sample is removed. However, if the test sample is to be removed from the bulk sample with quartering or a manual device such as a cup or scoop, then the bulk sample should be blended before the test sample is removed.

If the test sample is a granular product such as shelled corn or nuts, then the test sample should not be further reduced in size before grinding the sample in the sample preparation step. As the test sample becomes smaller, the uncertainty associated with estimating the true lot mycotoxin concentration increases. As shown later, the size of the test sample put through the grinder should be as large as possible. Recommended sample sizes for various commodities are shown in Table 6.1.

SAMPLE PREPARATION

Once a sample has been taken from the lot, the test sample must be prepared for mycotoxin quantification. Because it is not practical to extract the mycotoxin from a large test sample, it is usually extracted from a much smaller portion of product (subsample) taken from the test sample. If the commodity is a granular product such as shelled corn, it is essential that the entire test sample be comminuted in a suitable mill before a subsample is removed from the test sample (10, 11). Removing a subsample of whole seed from the test sample before the comminution process would eliminate the benefits associated with the larger size test sample of granular product. After the test sample has been comminuted, a subsample is removed for mycotoxin extraction.

Table 6.1. Product sample sizes used by the U.S. Food and Drug Administration

Product	Description	Package Type	Lot Size	# of Sample Units	Unit Size (kg)	Sample (kg)
Peanut butter	smooth	consumer bulk		24 12	0.23 0.45	5.44 5.44
Peanuts	crunchy butter, raw, roasted, ground topping	consumer & bulk		48	0.45	21.77
Treenuts	inshell, shelled, slices or flour paste	consumer & bulk		10 50 12	0.45 0.45 0.45	4.54 22.68 5.44
Brazil nuts	inshell in import status	bulk	<200 bags 201-800 bags 801-2000 bags	20 40 60	0.45 0.45 0.45	9.07 18.14 27.22
Pistachio nuts	inshell in import status	bulk	34,019 kg <34,019 kg	20% of units 20% of units		22.68 11.34
Corn	shelled, meal, flour, gritts	consumer & bulk		10	0.45	4.54
Cottonseed		bulk		15	1.81	27.22
Oilseed meals	peanut, cottonseed	bulk		20	0.45	9.07
Edible seeds	pumpkin, melon, sesame, etc	bulk		50	0.45	22.68
Ginger root	dried, whole ground	bulk consumer	"n" units	$\sqrt{n}$ 10	10x0.03	6.80 4.54
Milk	whole, low fat, skim	bulk consumer		10	1	4.54 4.54
Small grains	sorghum, wheat, barley, etc	bulk		10	0.45	4.54
Dried fruit	i.e. figs	consumer & bulk		50	0.45	22.68
Mixtures	-commodity particles large -commodity particles finely ground	consumer & bulk		50 10	0.45 0.45	22.68 4.54

Grinders should be used that reduce the particle size of the seed in the test sample to the smallest size possible. Grinders that produce small particles produce a more homogeneous test sample. As a result, the mycotoxin concentration of the subsample will more nearly reflect the true mycotoxin concentration of the test sample. Some grinders such as the Romer mill (12) and the USDA peanut mill (13) are designed to automatically provide a subsample during the grinding process. If the mill does not provide a subsample, the subsample can be obtained with a riffle divider. If the subsample is obtained with a manual device such as a scoop, the comminuted test sample should be blended before a subsample is scooped out.

Usually there is no sample preparation step associated with samples of nongranular products such as liquids (milk) or paste (peanut butter). A small portion of the sample may have to be removed for mycotoxin analysis because the entire sample cannot be analyzed. However, it is important to blend or mix liquid samples and paste samples before a small portion is removed for mycotoxin analysis.

Subsample sizes vary but usually are on the order of 25–1000 g depending on particle size. The smaller the particle size, the smaller the subsample size can be without increasing error or uncertainty.

ANALYTICAL QUANTIFICATION

Once a subsample is removed from the ground test sample, the mycotoxin is extracted from the comminuted subsample with solvents. Before the mycotoxin can be quantified in the solvent extract, analytical methods must be applied, which usually consist of several steps related to removing interfering compounds (i.e., oils) and concentrating the mycotoxins for quantification. These steps include centrifugation, filtration, drying, and dilution (14–16). Three types of methods can be used to quantify the mycotoxin extracted from the subsample: thin-layer chromatography, enzyme-linked immunosorbent assay methods that use antibody technology, and liquid chromatography (16). The FAO/WHO recommends using only analytical methods that have been collaboratively studied and that meet certain performance standards.

The mycotoxin measured from the solvent extract is expressed as a concentration: the ratio of the weight of the mycotoxin to the weight of the product in the subsample, or grams of mycotoxin per gram of product.

ACCEPT/REJECT LIMIT

Once the mycotoxin concentration is quantified, the sample value is compared with an accept/reject limit. The accept/reject limit is a predefined threshold value, usually equal to a legal limit used in regulatory applications. If the sample mycotoxin value is less than or equal to the accept/reject limit, the lot is accepted. Otherwise the lot is rejected. When lots are inspected by regulatory agencies, the accept/reject limit is usually set equal to the legal limit. However, manufacturers of consumer-ready products often use an accept/reject limit less than the legal limit to reduce the chances that consumer-ready products will be found by regulatory agencies with mycotoxin concentrations above the legal limit. Often, private industry uses an accept/reject limit that is about half the legal limit.

Many countries agree on the need to establish legal limits, but they often disagree on the value of the limit. A FAO survey in 1995 (17) showed that some countries have aflatoxin legal limits based on aflatoxin B_1 only and some countries use total aflatoxins ($B_1 + B_2 + G_1 + G_2$) and these legal limits vary widely. The Codex Committee on Food Additives and Contaminants has established a standard aflatoxin limit for peanuts at 15 µg/kg total aflatoxin for peanuts traded on the international market (6). This limit does not infringe on any nation's internal limits.

RISKS

Even when accepted sampling, sample preparation, and analytical procedures are used, there are errors (the term error is used to denote variability) associated with each of the above steps of the mycotoxin test procedure. Because of these errors, the true mycotoxin concentration in the lot cannot be determined with 100% certainty by measuring the mycotoxin concentration in a test sample taken from the lot. For example, 10 replicated aflatoxin test results from each of 12 contaminated shelled peanut lots are shown in Table 6.2 (18). For each sample, the mycotoxin test procedure consisted of (i) comminuting a 5.45-kg sample of peanut kernels in a USDA subsampling mill developed by

Dickens and Satterwhite (13, 19), (ii) removing a 280-g subsample from the comminuted test sample, (iii) extracting aflatoxins from a 280-g subsample as described by AOAC *Method II* (14), and (iv) quantifying the aflatoxins densitometrically by thin-layer chromatography. The 10 aflatoxin test results from each lot are ranked from low to high to demonstrate several important characteristics about replicated aflatoxin test results taken from the same contaminated lot.

First, the wide range among replicated test results from the same lot reflects the large variability associated with estimating the true mycotoxin content of a bulk lot. In Table 6.2, the variability is described by the standard deviation (SD) and the coefficients of variation (CV). Standard deviation is a measure of the dispersion of the 10 individual sample values (xi) about the mean (x) of the 10 sample values and can be calculated by Equation 2.

$$SD = [\Sigma(x_i - x)^2/(n - 1)]^{1/2} \qquad (2)$$

Coefficient of variation, expressed as a percent, is a measure of variability relative to the mean and can be defined as

$$CV = 100 \times (SD/x) \qquad (3)$$

The maximum test result can be four to five times the lot concentration (the average of the 10 test results is the best estimate of the lot concentration). Second, the amount of variation among the 10 test results appears to be a function of the lot concentration. As the lot concentration increases, the SD among test results increases, but the SD relative to the lot mean, as measured by CV, decreases. Third, the distribution of the 10 test results for each lot in Table 6.2 is not always symmetrical about the lot concentration. The distributions are positively skewed, which means that more than half the sample test results are below the true lot concentration. However, the distribution of sample test results becomes more symmetrical as the lot concentration increases. This skewness can be observed by counting the number of aflatoxin test results above and below the lot concentration in Table 6.2. If a single sample is tested from a contaminated lot, there is more than a 50% chance the sample test result will be lower than the true lot concentration. The skewness is greater for small sample sizes, and the distribution becomes more symmetrical as sample size increases (20).

Table 6.2. Distribution of aflatoxin test results for ten 5.4-kg samples from each of six lots of shelled peanuts[a,b]

Lot No.	Sample test result (µg/kg)										Mean (µg/kg)	SD[c] (µg/kg)	CV[d] (%) (µg/kg)
1	0	0	0	0	2	4	8	14	28	43	10	15	150
2	0	0	0	0	3	13	19	41	43	69	19	24	126
3	0	6	6	8	10	50	60	62	66	130	40	42	105
4	5	12	56	66	70	92	98	132	141	164	84	53	63
5	18	50	53	72	82	108	112	127	182	191	100	56	56
6	29	37	41	71	95	117	168	174	183	197	111	66	59

[a]From Whitaker et al. (1972).
[b]Aflatoxin test results are order by aflatoxin ppb.
[c]SD = standard deviation.
[d]CV = coefficient of variation = SD x 100/mean.

The variability shown in Table 6.2 is the sum of the variability associated with each step of the mycotoxin test procedure. As shown in Figure 6.3, the total variability (VT; using variance as the statistical measure of variability) associated with a mycotoxin test procedure is equal to the sum of the sampling (VS), sample preparation (VSS), and analytical variances (VA) associated with each step of the mycotoxin test procedure.

$$VT = VS + VSS + VA \qquad (4)$$

Examples of the magnitude of this variability associated with each step of a mycotoxin test procedure (Equation 4) are presented in the sections below. For example, the expected total variance associated with testing a shelled corn lot at 20 µg/kg when using a 0.91-kg sample, grinding the test sample in a Romer mill, taking a 50-g subsample from a comminuted sample, and using an immunoassay analytical method for quantification can be estimated by summing the variances associated with each step of the aflatoxin test procedure (Equation 4).

$$VT = 86.9 + 5 + 1.9 = 93.8 \qquad\qquad (5)$$

The variance, SD, and CV associated with the total aflatoxin test procedure described above are 93.8, 9.7, and 48.4%, respectively. The sampling, subsampling, and analytical variances account for 92.6, 5.3, and 2.1% of the total mycotoxin testing variance, respectively.

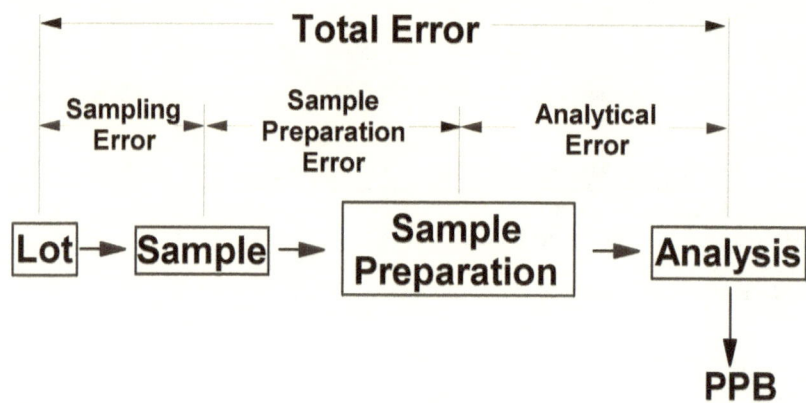

Figure 6.3. Types of errors associated with mycotoxin testing

As the preceding example demonstrates, the sampling step accounts for most of the variability (uncertainty) associated with the total variability of a mycotoxin test procedure. Studies by researchers on a wide variety of agricultural products (peanuts, cottonseed, shelled corn, wheat, barley, and pistachio nuts) indicate that a very small percentage of the kernels in the lot are contaminated and the concentration on a single kernel may be extremely high (21–30). For example, research has shown that aflatoxin in raw shelled peanuts may contaminate only 1 seed in 1000, or 0.1%. Also, Cucullu *et al.*, (31, 32) reported aflatoxin concentrations in excess of 1,000,000 μg/kg for an individual peanut kernel and 5,000,000 μg/kg for an individual cottonseed. Shotwell *et al.*, (33) reported finding over 400,000 µg/kg of aflatoxin in an individual corn kernel.

Because of this extreme mycotoxin distribution among seed in a contaminated lot, it is easy to miss the contaminated seed with a small sample and underestimate the true lot concentration. On the other hand, if the test sample contains one or more highly contaminated seeds, then it will measure high and overestimate the true mycotoxin contamination in the lot. Even with proper sample selection techniques, the variation among test sample concentrations is large because of the mycotoxin distribution described above.

REDUCING VARIABILITY OF A MYCOTOXIN TEST ROCEDURE

The only way to achieve a more precise estimate of the true lot concentration is to reduce the total variability of the test procedure. The variability of the test procedure can be reduced by reducing the variability associated with each step of the mycotoxin test procedure. Increasing the size of the sample can reduce the sampling variability. The sample preparation variability can be reduced by increasing the size of the subsample and/or by increasing the degree of comminution (increasing the number of particles per unit mass in the subsample). The analytical variance can be reduced by increasing the number of aliquots quantified by the analytical method and/or by using a more precise quantification method (i.e., liquid chromatography instead of thin-layer chromatography). If the variability associated with one or more of these steps can be reduced, then the total variability associated with a mycotoxin test result can be reduced. The range of mycotoxin test results associated with any size sample and subsample, and the number of analyses about the lot concentration M can be estimated from the SD (square root of the total variance) associated with the mycotoxin test procedure. About 95% of all test results will fall between a low of ($M - 1.96 \times$ SD) and a high of ($M + 1.96 \times$ SD). The two expressions are valid only for a normal distribution where test results are symmetrical about the mean. The distribution among aflatoxin test results is usually skewed but approaches a symmetrical distribution as sample size becomes large.

The effect of increasing sample size on the range of test results when a contaminated lot of shelled corn that has 20 µg/kg of aflatoxin is being tested is shown in Table 6.3. The range does not decrease at a constant rate as sample size increases. For example, doubling sample size has a greater effect on decreasing the range at small sample sizes than at large sample sizes. This characteristic suggests that increasing sample size beyond a certain point may not be the best use of resources and that increasing subsample size or number of analyses may be a better use of resources in reducing the range of test results once sample size has become significantly large. Increasing the sample size by a factor of five, from 0.91 to 4.54 kg, will cut the sampling variance in Equation 5 by a factor of five to 17.4. The total variance with the 4.45-kg sample now becomes

$$VT = 17.4 + 5.0 + 1.9 = 24.3 \qquad\qquad (6)$$

The variance, SD, and CV associated with the total testing procedure have been reduced to 24.3, 4.9, and 24.6%, respectively.

Table 6.3. Range of aflatoxin estimates for 95% confidence limits when testing a contaminated lot of shelled corn with 20 µg/kg aflatoxin using different sample sizes

Sample size (kg)	Standard deviation[a] (µg/kg)	Low[b] (µg/kg)	High[c] (µg/kg)
1	9.2	2.0	38.0
2	6.8	6.7	33.3
4	5.2	9.8	30.2
8	4.1	12.0	28.0
16	3.4	13.3	26.7
32	3.1	13.9	26.1

[a]Standard deviation reflects sample size shown in table plus a 50-g subsample that will pass a no. 20 screen and immunoassay analytical method. Sample preparation plus analytical standard = 2.6 and is constant for all sample sizes.
[b]Low = 20 - 1.96 (standard deviation).
[c]High = 20 + 1.96 (standard deviation).

-DESIGNING MYCOTOXIN SAMPLING PLANS

Because of the large variability among test results, two types of mistakes are associated with any mycotoxin-sampling program. First, good lots (lots with a concentration less than or equal to the legal limit) will test bad and be rejected by the sampling program. This type of mistake is often called the seller's risk, because these lots are rejected at an unnecessary cost to the seller of the product. Second, bad lots (lots with a concentration greater than the legal limit) will test good and be accepted by the sampling program. This type of mistake is called the buyer's risk, because contaminated lots will be processed into feed or food, causing possible health problems and/or economic loss to the buyer of the product. To maintain an effective regulatory and/or quality control program, the above risks associated with a sampling design must be evaluated. Based on these evaluations, the costs and benefits (benefits refers to removal of mycotoxin-contaminated lots) associated with a sampling program can be evaluated.

A lot is termed bad when the sample test result X is above some predefined accept/reject limit X_c, and the lot is termed good when X is less than or equal to X_c. Although X_c is usually equal to the legal limit M_c, it can be greater than or less than M_c. For a given sample design, lots with mycotoxin concentration M will be accepted with a certain probability $P(M) = \text{prob}(X < X_c | M)$ by the sampling plan. A plot of $P(M)$ versus M is called an operating characteristic (OC) curve. Figure 6.4 depicts the general shape of an OC curve. As M approaches 0, $P(M)$ approaches 1; as M becomes large, $P(M)$ approaches zero. The shape of the OC curve is uniquely defined for a particular sampling plan design with designated values of sample size, degree of comminution, subsample size, type of analytical method, number of analyses, and the accept/reject limit X_c.

For a given sampling plan, the OC curve indicates the magnitudes of the buyer's and seller's risk. When M_c is defined as the legal limit or the maximum lot concentration acceptable, lots with $M > M_c$ are bad and lots with $M \leq M_c$ are good. In Figure 6.4, the area under the OC curve for $M > M_c$ represents the buyer's risk (bad lots accepted), and the area above the OC curve for $M < M_c$ represents the seller's risk (good lots rejected) for a particular sampling plan.

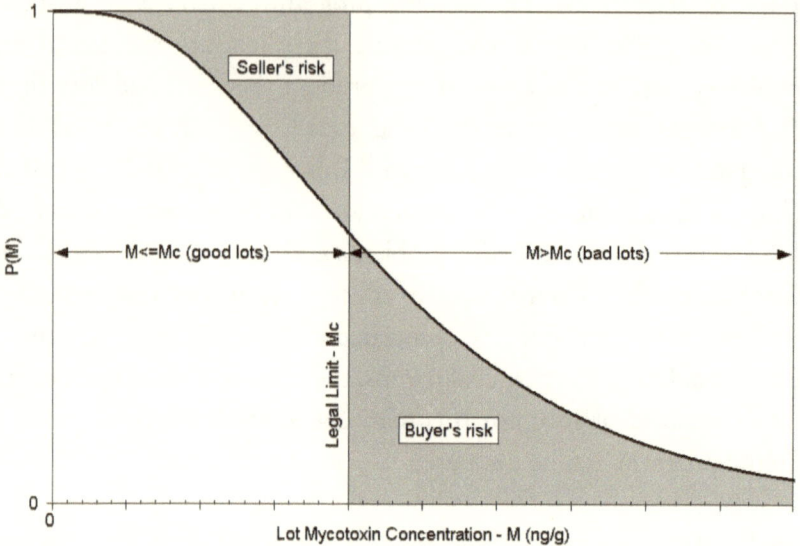

Figure 6.4. Typical shape of an operating characteristic curve used to calculate the buyer's and seller's associated with a sampling plan.

Because the shape of the OC curve is uniquely defined by the sample size, degree of comminution, subsample size, number of analyses, and the accept/reject limit X_c, these parameters can be used to reduce the buyer's and seller's risks associated with a sampling plan. The effect of increasing sample size on the shape of the OC curve when testing shelled corn lots for aflatoxin is shown in Figure 6.5, where the accept/reject limit is equal to the legal limit of 20 µg/kg. As sample size increases from 0.91 to 9.07 kg, the slope of the OC curve about the legal limit increases, forcing the two areas associated with each risk to decrease. As a result, increasing the size of the sample decreases both the buyer's and the seller's risk. The same effect can be obtained by increasing the degree of sample comminution, the subsample size, or the number of analyses.

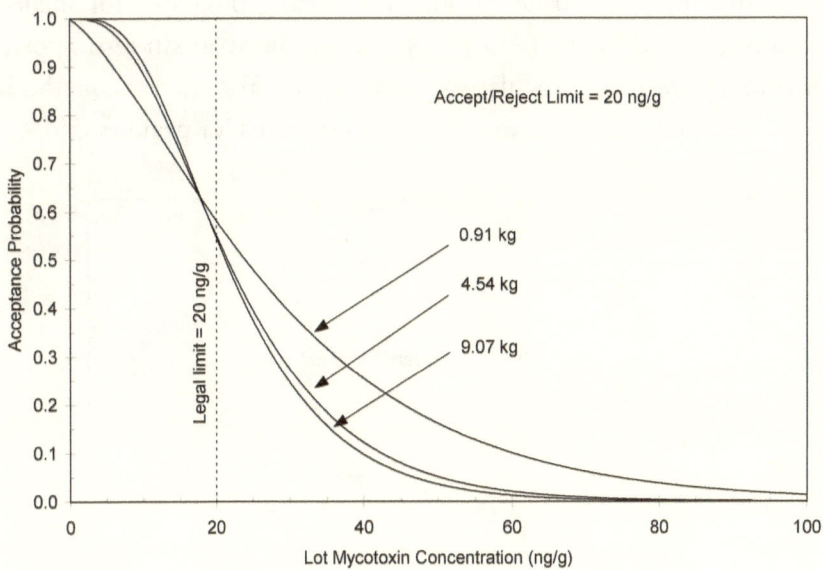

Figure 6.5. Effect of sample size on the buyer's and seller's risks associated with sampling shelled corn for aflatoxin.

The effect of changing the accept/reject limit, relative to the legal limit, on the two risks when testing shelled corn lots for aflatoxin is shown in Figure 6.6. If the legal limit is assumed to be 20 µg/kg, then changing X_c to a value less than 20 µg/kg shifts the OC curve to the left. Compared with the sampling plan where X_c = 20, the buyer's risk decreases, but the seller's risk increases. If X_c becomes larger than 20, the OC curve shifts to the right. As a result, the seller's risk decreases but the buyer's risk increases. Changing the accept/reject limit relative to the legal limit can reduce only one of the two risks, because reducing one risk automatically increases the other risk.

Methods have been developed to predict the seller's and buyer's risks, the total number of lots accepted and rejected, the amount of mycotoxin in the accepted and rejected lots, and the costs associated with a mycotoxin inspection program for several commodities (34–36). These methods have been used by the USDA, Agricultural Marketing Service and the peanut industry to design aflatoxin-testing programs for shelled peanuts (37) and by the FAO (34) to design the aflatoxin-testing plan for corn and peanuts. Regulatory agencies have also used the methods to evaluate existing aflatoxin inspection programs for peanuts (35).

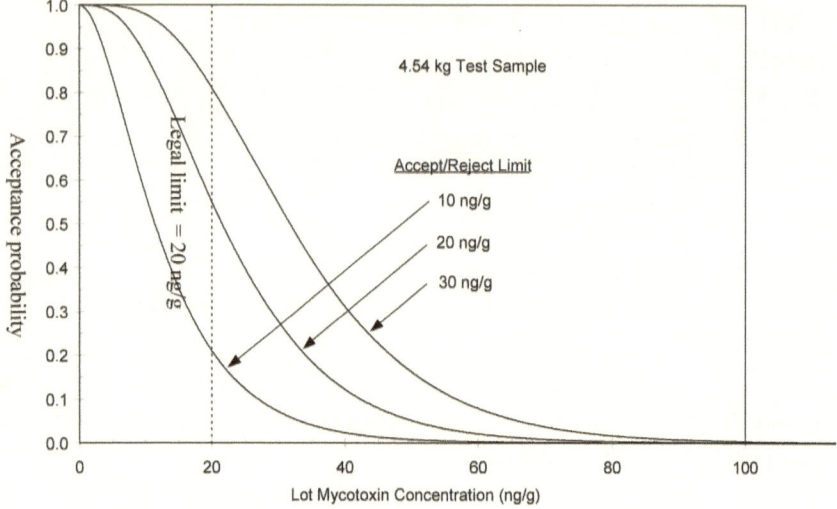

Figure 6.6. Effect of accept/reject limit on the buyer's and seller's risks associated with sampling shelled corn for aflatoxin.

SUMMARY

A mycotoxin-sampling plan is defined by a mycotoxin test procedure and a defined accept/reject limit. A mycotoxin test procedure is a complicated process and generally consists of several steps: (i) a sample is taken from the lot, (ii) the sample is ground (comminuted) in a mill to reduce particle size, (iii) a subsample is removed from the comminuted sample, and (iv) the mycotoxin is extracted from the comminuted subsample and quantified. Even when accepted test procedures are used, there is variability associated with each step of the mycotoxin test procedure. Because of this variability, the true mycotoxin concentration in a lot cannot be determined with 100% certainty by measuring the mycotoxin concentration in a sample taken from the lot. Even when the sample is correctly selected, there will be variability associated with the mycotoxin test procedure. The variability for each step of the mycotoxin test procedure, as measured by the variance statistic, increases with mycotoxin concentration. For small sample sizes, sampling is usually the largest source of variability associated with the mycotoxin test procedure. Sampling variability is large because a small percentage of kernels are contaminated and the amount of contamination on a single seed can be very large. Increasing sample size, degree of sample comminution, subsample size, and the number of aliquots quantified can reduce the variability associated with a mycotoxin test procedure. Reducing the variability of the mycotoxin test procedure reduces the number of lots misclassified by the sampling plan.

REFERENCES

(1)   Gy, P.M. (1982) *Sampling of Particulate Materials*, Elsevier, Amsterdam, The Netherlands

(2)   William, P.C. (1991) in *Mycotoxins and Animal Foods*. Smith, J.E. & Henderson, R.S. (Eds), CRC Press, Boca Raton, FL, 721

(3)   Shotwell, O.L., Goulden, M.L., Botast, R.J., & Hasseltine, C.W. (1975) *Cereal Chem.* **52**, 687

(4)   Bauwin, G.R., & Ryan, H.L. (1982) in *Storage of Cereal Grains and Their Products*. Vol. 5, Christensen, C.M. (Ed), American Association of Cereal Chemists, St. Paul, MN, 115

(5)   Hurburgh, C.R., & Bern, C.J. (1983) *Trans. Am. Soc. Agric. Eng.* **26**, 930

(6)   FAO (2001) *Proposed Draft Revised Sampling Plan for Total Aflatoxin in Peanuts Intended for Further Processing*, Joint FAO/WHO Food Standards Program, Codex Alimentarius Commission, 24th Session, Geneva, Switzerland, July 2–7, 2001, FAO/WHO Joint Office, Food and Agriculture Organization, Rome, Italy, 276

(7)   USDA (1975) *Inspectors Instructions*, Agricultural Marketing Service, United States Department of Agriculture, Washington, DC

(8)   Parker, P.E., Bauwin, G.R., & Ryan, H.L. (1982) in *Storage of Cereal Grains and Their Products*, Christensen, C.M. (Ed), American Association of Cereal Chemists, St. Paul, MN

(9)   Whitaker, T.B., & Dowell, F.E. (1995) in *Advances in Peanut Science*, H.E. Pattee & H.T. Stalker (Eds) American Peanut Research and Education Society, Stillwater, OK

(10)   Dickens, J.W., & Whitaker, T.B. (1982) in *Environmental Carcinogens—Selected Methods of Analysis: Some Mycotoxins*. Vol. 5, H. Egan, L. Stoloff, P. Scott, M. Costegnaro, I.K. O'Neill, & H. Bartsch (Eds), ARC, Lyon, France

(11)   Campbell, A.D., Whitaker, T.B., Pohland, A.E., Dickens, J.W., & Park, D.L. (1986) *Pure Appl. Chem.* **58**, 305

(12)   Malone, B. (2000) in *Mycotoxin Protocols*. M.W. Trucksess & A.E. Pohland (Eds), Humana Press, Totowa, NJ

(13)   Dickens, J.W., & Satterwhite, J.B. (1969) *Food Technol.* **23**, 90

(14)   AOAC (1990) *Official Methods of Analysis* 1990, 15th Ed., AOAC INTERNATIONAL, Arlington, VA, Method II

(15)   Nesheim, S. (1979) *Methods of Aflatoxin Analysis*, NBS Special Publication (US) No. 519, NBS, Washington DC

(16)   Steyn, P.S., Thiel, P.S., & Trinder, D.W. (1991) in *Mycotoxins and Animal Foods*. J.E. Smith & R.S. Henderson (Eds), CRC Press, Boca Raton, FL

(17)   FAO (1995) *Worldwide Regulations for Mycotoxins 1995*. FAO Food and Nutrition Paper 64, Food and Agriculture Organization, Rome, Italy

(18)   Whitaker, T.B., Dickens, J.W., & Monroe, R.J. (1972) *J. Am. Oil Chem. Soc.* **49**, 590

(19)   Dickens, J.W., Whitaker, T.B., Monroe, R.J., & Weaver, J.N. (1979) *J. Am. Oil Chem.* Soc. **56**, 842

(20)   Remington, R.D., & Schrok, M.A. (1970) *Statistics and Applications to the Biological and Health Sciences*, Prentice-Hall, Englewood Cliffs, NJ

(21)   Whitaker, T.B., Dickens, J.W., & Monroe, R.J. (1974) *J. Am. Oil Chem. Soc.* **51**, 214

(22)   Whitaker, T.B., Whitten, M.E., & Monroe, R.J. (1976) *J. Am. Oil Chem. Soc.* **53**, 502

(23)   Whitaker, T.B., Dickens, J.W., & Monroe, R.J. (1979) *J. Am. Oil Chem. Soc.* **56**, 789

(24)   Whitaker, T.B., Dowell, F.E., Hagler, W.M., Jr., Giesbrecht, F.G., & Wu, J. (1993) *J. AOAC Int.* **77**, 107

(25)   Schatzki, T.F. (1995) *J. Agric. Food. Chem.* **43**, 1561

(26)   Schatzki, T.F. (1995) *J. Agric. Food. Chem.* **43**, 1566

(27)   Whitaker, T.B., Truckess, M., Johansson, A., Giesbrecht, F.G., Hagler, W.M., Jr., & Bowman, D.T. (1998) *J. AOAC Int.* **81**, 1162

(28)   Hart, L.P., & Schabenberger, O. (1998) *Plant Dis.* **82**, 625

(29)   Johansson, A.S., Whitaker, T.B., Hagler, W.M., Jr., Giesbrecht, F.G., Young, J.H., & Bowman, D.T. (2000) *J. AOAC Int.* **83**, 1264

(30)   Whitaker, T.B., Hagler, W.M., Jr., Giesbrecht, F.G., & Johansson, A.S. (2000) *J. AOAC Int.* **83**, 1285

(31)   Cucullu, A.F., Lee, L.S., Mayne, R.Y., & Goldblatt, L.A. (1986) *J. Am. Oil Chem. Soc.* **43**, 89

(32)   Cucullu, A.F., Lee, L.S., & Pons, W.A. (1977) *J. Am. Oil Chem. Soc.* **54**, 235A

(33)   Shotwell, O.L., Goulden, M.L., & Hessletine, C.W. (1974) *Cereal Chem.*, **51**, 492

(34)   FAO (1993) *Sampling Plans for Aflatoxin Analysis in Peanuts and Corn*. FAO Food and Nutrition Paper 55, Food and Agriculture Organization, Rome, Italy

(35)   Whitaker, T.B., & Dickens, J.W. (1979) *Peanut Sci.* **6**, 7

(36)   Johansson, A.S., Whitaker, T.B., Hagler, W.M., Jr., Giesbrecht, F.G., & Young, J.H. (2000) *J. AOAC Int.* **83**, 1279

(37)   Whitaker, T.B., Springer, J., Defize, P.R., deKoe, W.J., & Coker, R. (1995) *J. AOAC Int.* **78**, 1010

7. SAMPLING VILLAGE CORN FOR AFLATOXIN ANALYSIS: PRACTICAL ASPECTS

Henry Njapau

Center for Food Safety and Applied Nutrition
Food and Drug Administration
5100 Paint Branch Parkway
College Park, Maryland 20740, USA

Environmental conditions that accelerate mycotoxin formation in agricultural commodities are most commonly encountered in tropical regions that are also home to the world's poorest communities. These communities depend on subsistence agriculture, producing small quantities of food that are barely sufficient to last a household from one growing season to another. Food commodities are stored in various forms and structures that may include, but not limited to, rooftops, kitchen ceilings, tree branches or open cribs for ears of corn, and clay pots, sealed mud cribs or sacks for shelled corn and peanuts.

This overview presents situations frequently encountered when collecting samples of on-farm or stored agricultural commodities from rural communities in most developing countries for aflatoxin analysis. Often, the characteristics of the storage facilities render the use of established probability-based official sampling plans difficult. Instead, an investigator has to modify an established plan or formulate an alternative sampling protocol that is appropriate for the specific situation. Some of the factors that dictate or influence the manner in which the sample is collected are outlined and alternative sampling approaches are presented. For a detailed discussion of specific official sampling plans and the principles upon which their establishment is based, the reader is referred to chapter 6 of this book, a Food and Agriculture Organization (FAO) manual (1) and other resources (2-7) from which material used in some sections of this discussion was abstracted.

THE SAMPLE

A sample is generally a small portion obtained from a larger quantity of material or population (lot) and used to ascertain attributes of the lot from which it is drawn (Table 7.1). Occasionally, all the material available is collected as a sample. The latter situation may be encountered when collecting samples of ready-to-eat meals (plate foods) or serving-size portions of street snack foods. Sampling could be uniform (for products of homogeneous quality), selective (targeted to e.g. detect lowest quality components of a lot) or random (for unevenly occurring attributes). When a sample is collected from a lot where the attribute of interest is not uniformly distributed, random sampling is the most appropriate technique. Random sampling involves collecting several small portions (increments) from different areas/sections of the lot, in a manner that ensures that all individual components of the lot have an equal chance of becoming part of the sample. The incremental portions are combined into one aggregate or composite sample that *represents* the lot within set precision bounds. Random samples make possible the use of probability theory to estimate how different an attribute measured in the sample is likely to be from the true average content of that attribute in the lot. Reference to the term sampling in the remainder of this discussion implies random sampling or adaptations thereof.

The number of increments that constitute the composite sample is not a critical factor where the attribute of interest is uniformly (homogenously) distributed in the lot. For instance, since oxygen in the air we breathe is expected be homogenously distributed within a 100 m² space at similar elevation, a measurement of the oxygen tension in one area of the designated space can reasonably be assumed to represent the oxygen tension in the entire space. Similarly, the lactose content of a well stirred

Table 7.1. Definitions of terms common in sampling literature

Term	Definition	Scope
Lot	Indentifiable commodity produced/stored under uniform conditions.	Can be applied to crops in the field, commodities in storage (bulk or family-size), a batch off a production line, or a consignment.
Consignment	Any commodity, material or group of items delivered at the same time.	A shipload, trainload, single truck, or ox-drawn cart of corn. Multiple deliveries within less than a few days, under similar conditions and from a single source can constitute one consignment.
Sample	Portion drawn from a lot, consignment or population that provides general or specific information about the lot.	Depending on the sampling plan and characteristics of the specific analyte, the sample may be a one adult size portion of a meal, a few ears of corn from a household food storage bin or 30 kg.
Primary or Incremental sample	Quantity or group of items collected at once from one place in a lot.	Constitutes all the grains in a single probe (trier), all the fuzzy cottonseed in a single corkscrew probe or a single handful of material (often 200-300 g).
Composite or aggregate sample	Mixture of multiple primary or incremental samples.	Combination of several small increments (primary samples) are drawn from one lot.
Final/laboratory sample (test sample)	Composite or portion of the composite sample designated for grinding or laboratory analysis.	May constitute a single collection (e.g. plate food), entire composite sample or suitable portion of composite sample obtained by an appropriate quantity reduction process.

5,000 L tank of milk is not expected to vary substantially between different sections of the tank. The lactose content of one sample collected from any portion of the tank should represent the lactose content of the entire tank. Unlike the oxygen and lactose examples, the distribution of aflatoxin in a lot of corn or among individual kernels on an ear (cob) of corn has been shown to be extremely uneven (heterogeneous) (4, 8). Only a few kernels on an ear may be highly contaminated with aflatoxin while the rest of the kernels contain little or no aflatoxin at all. It has further been demonstrated that one half of a peanut kernel may contain a large amount of aflatoxin while the other half is aflatoxin free (9). Accordingly, the average aflatoxin content of a household small farm, food store, or a single 90 kg sack of corn may be a result of contamination occurring in only a few ears or kernels, a situation similar to the *Pareto Principle* or *80-20* rule (10), which states that 80% of a problem is caused by 20% of the causes. Therefore, a test result from a single small portion collected from one section of a farm, storage facility, or sack of corn is likely to misrepresent the true average aflatoxin concentration of the lot if a few highly contaminated kernels were either included in or excluded from the small portion. The ideal situation would be to either analyze the entire lot or several small portions whose results are then averaged. The cost of both options would be prohibitive. Instead, the several small portions that are collected are combined into one composite sample that is analyzed. Current official sampling plans have demonstrated that analyzing the composite sample yields results of acceptable accuracy.

FORMULATING A SAMPLING PROTOCOL

When developing an official sampling plan, several factors including, the type and characteristics of the commodity to be tested, aflatoxin contamination distribution, performance of the specified analytical procedure and an accept/reject level, are considered. A typical sampling plan will thus specify lot or sublot sizes, the number and size of increments, the minimum composite sample size, a sample reduction and/or preparation method, an analytical method and the accept/reject guideline level for a particular commodity. Official sampling plans are designed for optimal performance while minimizing risks to both consumers and processors of agricultural commodities (11). Since the plans are principally for commodities in national or international commerce, the number of increments and ultimate size of the composite sample usually constitute a very small percentage of the lot. Hence, it is not uncommon for various plans to stipulate samples sizes of 20 to 30 kg (2, 3, 7, and 12). In general, the larger the number of kernels in the sample the higher the precision of the estimation.

Rural communities in developing countries produce relatively small quantities of a particular food commodity hence lot sizes are likely to be much smaller than sizes encountered in commercial channels. A household's foodstore may in times of food scarcity, contain less than 30 kg of grain. If samples of foodstuffs should be collected from these localities in order to assess aflatoxin contamination, an appropriate sampling protocol should be formulated prior to the execution of the sampling exercise. Local authorities or community leaders can be a source of information on the size of household farms, common types of storage facilities and whether the crop is in the field or in storage. Although it is desirable that one method of collecting the samples is utilized throughout, adjustments to the scheme may be necessary from time to time as hitherto unanticipated situations are encountered. Consequently, the relationship between the sample and the lot may deviate from the desired similarity with the adjustments to the sampling scheme. Nonetheless, the principle of collecting incremental portions to ensure that the sample represents the lot from which it is drawn must be adhered to as much as is logistically possible. The variations must be documented in as much detail as possible. Moreover, the sample collectors must be mindful of the social and economic implications of collecting the required sample size, particularly if an equivalent replacement quantity is not provided to the household.

COLLECTING THE SAMPLE

The sampling protocol should describe the manner in which the sample should be collected (e.g. hand grabs from the top, middle and bottom of a sack or use of triers/probes) and the sampling scheme or pattern [e.g., x kg per every 2 hectares of a corn farm; one-tenth of the number of corn ears in a crib; every y^{th} bag being loaded onto a truck; probe z number of bags on each face of a stack of 5,000 metric tons (MT) of corn]. The U.S. Grain Inspection, Packers and Stockyards Administration (GIPSA) (6) recommends systematic collection of material throughout a lot that yields an approximate random sample. GIPSA does not permit hand grabs or cup scoops as a way of collecting samples. The reasons for the sampling scheme must be made clear to everyone involved in the exercise, including the household whose food is being sampled. Emphasizing the fact that only a few kernels on an ear or in a 50-90 kg sack of corn may contain aflatoxin, and that collecting only those kernels or missing them totally will produce an erroneous test result is important.

Smallholder Village Farms

When collecting samples from crops in the field it is important to realize that, in addition to the uneven occurrence of aflatoxin among kernels on an ear, there may be variation within a single field and/or between fields. And, where growth of the stalks is of uneven height, it is not farfetched to assume potential differences due to proximity to the soil surface. Therefore, aflatoxin levels determined from field samples may be less representative (of the average contamination of the farm) than those from harvested and shelled corn from the same farm. However, when samples must be collected from the field, efforts should be made to ensure that all ears of corn have an equal chance or being included in the sample. An erroneous estimate would be likely if corn ears constituting the sample were largely those heavily contaminated or those relatively free of aflatoxin.

Small-scale village farms may vary from one to several hectares. The size of the composite sample of corn ears will similarly vary according to the size of the farm. There is a paucity of information regarding collecting samples from the field. It is perhaps an indication of either established or perceived large variability in estimates of aflatoxin contamination of a field of corn. Furthermore, corn in the field may be at a moisture content that could still permit fungal growth necessitating immediate drying to prevent aflatoxin build-up between collection and time of analysis. Below and in Table 7.2 are some suggestions for collecting corn samples from the field for aflatoxin analysis.

(1)   For fields of 1 hectare or less, it is recommended that:

   a)   The field is artificially subdivided into 10 x 10 m quadrants. One ear of corn is collected from a randomly selected point within each quadrant. This scheme will yield approximately 5 ears/acre.

Table 7.2. Examples of sampling specifications for shelled corn and raw shelled peanuts

Commodity	Country/ Institution	Form of commodity	Lot size	No. of Increments	Increment size (kg)	Aggregate sample size (kg)	Minimum laboratory Sample (kg)
	USFDA[a]	Bulk	–	10	0.45	4.5	1 x 4.5
		Sacks	10	13	0.20	2.6	1 x 5.0
Corn (shelled)	Fonsec[b]	Sacks	50	28	0.20	5.7	1 x 5.0
		Bulk	100MT	180	0.20	36.0	7 x 5.0
	EC[c]	Bulk	≤1MT	10	0.10	1.0	1 x 1.0
		Bulk	20-50MT	100	0.10	10	1 x 10
	USFDA	Bulk	(If 4.5 kg	10	0.45	4.5	1 x 4.5
	USFDA	Bulk	is + ve)	48	0.45	21.8	1 x 21.8
		Sacks	10	19	0.20	2.6	1 x 5.0
Peanuts (raw & shelled)	Fonseca	Sacks	50	42	0.20	5.7	1 x 5.0
		Bulk	100 MT	240	0.20	40	8 x 5.0
		Bulk	≤ 1 MT	10	0.30	3	1 x 3.0
	EC	Bulk	0.5-1 MT	30	0.30	9	1 x 3-12
		Bulk	15-125MT	100	0.30	10	1 x 30
	Codex[d]	Bulk	25MT	100	0.20	20	1 x 20

[a]United States of America Food and Drug Administration 2007
[b]Fonseca, 2002 (if composite sample is less than laboratory sample, resample same bags; if more, make seperate 5 kg samples)
[c]Commission Regulation (EC) No. 401/2006.
[d]CX/FAC 05/37/23, Annex 1, December 2004.

b) The ears are shelled, the kernels mixed thoroughly and the entire sample is ground if ≤ 5 kg. If the composite sample is > 5 kg, the sample is thoroughly mixed and the size reduced to ~ 5 kg (see sample processing).

(2) For medium size farms, 2 to 5 hectares, 3 ears are randomly collected per acre. The ears are shelled and a 5 kg sample is removed (see sample processing). Although farms larger than 5 hectares are less frequent, there may be instances where they may be encountered. In such instances the number of ears randomly collected per acre is reduced to two.

(3) In certain areas of Zambia, stalks of corn are often grouped (Figure 7.1A) to facilitate hand removal of ears from the stalks. Each group of stalks can be considered loci for collecting incremental samples of one or two ears from the periphery and the center, depending on the estimated number of stalks constituting the pile.

Courtesy: Masja Straetemans

Courtesy: Mt. Makulu Agricultural Research Station, Zambia

Figure 7.1. Harvesting and storage of corn. Corn stalks grouped during harvesting (A), traditional corn granary made from poles (B), an improved version of the traditional corn gruanary (C), and experimental *Ferrumbu* type shelled corn storage structures at the Central Agricultural Research Station, Mt. Makulu, Chilanga, Zambia (D). Note traditional mud crib in the rear and an improved larger version in the center of plate D.

Postharvest Commodities

Official sampling plans currently in use were developed for postharvest bulk commodities in commerce. An informal review of the plans shows minor variations from country to country or region to region, largely as a result of the extent of risk determined to be acceptable to consumers and food processors. Two unofficial sampling protocols and two official plans are referenced here because of their relevance to potential village sampling scenarios. Aspects of all four sampling protocols pertinent to small size lots and some elements of the GIPSA grain sampling procedures were used as a foundation for the recommendations given herein.

The unofficial sampling procedures are: a procedure for sampling shelled corn and shelled raw peanuts developed by Professor Fonseca, University of Sao Paulo, SP, Brazil ("Fonseca protocol") (2) and a sampling scheme by Hongsuwong published in FAO RAP Publication-1989, *Mycotoxin Prevention and Control in Foodgrains* ("Hongsuwong protocol"). The official plans are: the European Economic Commission (EC) (3) plan and the U.S. Food and Drug Administration (FDA) plan (12). The relevant aspects of the four sampling protocols are shown in Table 7.3.

Table 7.3. Proposed protocol for collecting samples of corn ears in the field, shelled loose grain corn, and ready-to-eat meals for aflatoxin analysis

Nature of commodity	Recommended sampling protocol
Field, 1 hectare or less	• Divide field into equal sections and collect one ear (cob) from each section and composite into one test sample (5 ears/acre).
Field, 2 to 5 hectares	• Divide field into approximate acre sections • Collect 3 ears (cobs) of corn from each section • Shell and reduce to a 5 kg test sample
Farm or field greater than 5 hectares (13 acres)	• Collect 2 ears/acre
Household storage. Open Crib with corn on ears	• Estimate total quantity and collect sample from accessible zones and composite. Collect ears to yield 1 kg/100 kg corn
Trucks and rail cars	• Collect 5-8 probes per wagon (see chapter 6 for details) and composite (minimum size 5 kg).
Grain in a ship	• May sample with pneumatic probes although not appropriate • Difficult to collect representative sample • Best when loading and/or offloading
Plate foods Street/market snacks	• Collect whole meal as single test sample • May collect snack portions as individual test samples or may combine several portions into one test sample depending on purpose.

The Fonseca publication is not the official sampling plan of the government of Brazil (Scussel, personal communication). Similarly, the Hongsuwong sampling scheme may not be a fully endorsed official FAO protocol. An FAO officially recommended protocol, a Codex Commission recommendation, is shown in Table 7.3. Both the Fonseca and Hongsuwong protocols are based on the use of the square root of the total quantity (lot) to calculate the number of increments hence the composite sample size. The Fonseca protocol stipulates 0.2 kg as the minimum incremental size and includes an additional lot size-based factor. The number of increments stipulated by the Fonseca protocol is determined using the equation $N = x\sqrt{nS}$, where 'N' is the number of increments, 'x' is a numeric lot size-based variable factor and 'nS' is the number of sacks constituting a lot of shelled corn or peanuts. For corn or peanuts in sacks, $x = 4$, and $x = 18$, when they are in large bulk form. The minimum laboratory sample size is 5 kg. The Hongsuwong protocol is for corn only and recommends collecting 3 increments (quantity not specified) from a minimum of 10 sacks for lots of 10 to 100 sacks. For larger lots, increments are collected from a square root of the total quantity. When the sacks are stacked (piled) the Hongsuwong protocol recommends collecting incremental samples of at least 0.5 kg from the sides and top of the pile covering no less than 5% of the sacks. For bulk corn (up to 500 MT) the Hongsuwong protocol stipulates 1 kg incremental samples. The minimum laboratory sample from bulk lots is 5 kg.

The EC plan for sampling bulk cereals and cereal products specifies a minimum of 10 kg composite (aggregate) sample from 100 x 0.1 kg incremental samples for lots of 20 to 50 MT. The entire 10 kg composite sample is analyzed as one laboratory sample. Larger lots are subdivided into 100 – 500 MT sublots from which the 10 kg composite sample is drawn. The EC plan for sampling cereals allows for smaller composite sample sizes, up to a minimum of 1 kg from at least 3 incremental samples, for lots of 1 MT or less. Similar gradations are allowed under the EC plan for sampling raw shelled peanuts. The stipulated minimum number of incremental portions and composite sample size for peanut samples under the EC plan are

100 and 30 kg, respectively, for ≥15 MT lots. For peanut lots of 100 kg or less a minimum of 10 increments and a 3 kg composite sample are required. The entire composite sample (30 kg or 3 kg) is analyzed i.e. constitutes a laboratory sample. The EC regulation also provides a formula for calculating the sampling frequency for commodities in individual containers such as sacks or packs from a retail outlet. The equation is given as:

$$SF = (Wt.\ of\ lot\ x\ Wt.\ of\ incremental\ sample) \div Wt.\ composite\ sample\ x\ Wt.\ of\ sack)$$

Where SF = frequency of sampling, i.e. every n^{th} sack from which an incremental sample must be taken.

For a composite sample of 10 kg; an incremental sample size of 0.1 kg and 90 kg sacks of corn, the equation yields a frequency of every 100th sack. This would require collecting 0.1 kg incremental samples from 100 sacks of a 10,000-sacks lot and from only 1 sack of a 100-sacks lot. The number of sacks from which the incremental samples are collected doubles if the incremental weight is increased to 0.2 kg. The equation yields acceptable sampling frequencies when applied to large but not small lots. In order to collect from at least 10 of a 100-sacks lot, the incremental weight would have to be 1 kg.

The FDA plan for shelled peanuts is a two-step process that begins with a 4.5 kg composite (and laboratory) sample constituted from 10 x 0.45 kg increments. If the 4.5 kg is determined to contain aflatoxin above the action level, a second 21.8 kg sample is collected and analyzed. Table 7.4 Shows comparisons of the Fonseca, Hongsuwong and EC plans.

Table 7.4. Sampling frequencies [number of increments per lot, (N)] and sample sizes for three (3) corn sampling schemes

Weight MT	No. of Sacks	Fonseca			Hongsuwong			EC			Mean
		(N)	Comp. sample size (kg)	Lab sample size (kg)	(N)	Comp. sample size (kg)	Lab sample size (kg)	(N)	Comp. sample size (kg)	Lab sample size (kg)	Lab sample size (kg)
0.900	10x90kg	13	2.6	1x5.0	1.0	5.0?[a]	1x5.0?	10	1.0	1x1.0	1-2
	18x50kg	17	3.4	1x5.0	10	5.0?	1x5.0?	10	1.0	1x1.0	
9.000	100x90kg	40	8.0	2x5.0	10	5.0?	1x5.0?	40	4.0	1x4.0	4-5
	180x50kg	54	10.8	2x5.0	14	7.0?	1x5.0?	40	4.0	1x4.0	
90.00	1000x90kg	127	25	5x5.0	32	16.0	1x5.0	100	10	1x10	5-10
	1800x50kg	170	34	7x5.0	42	21.0	1x5.0	100	10	1x10	

[a]The Hongsuwong protocol does not clearly specify incremental or sample size for lots of 10-100 sacks.

As earlier stated, it is not uncommon for small rural households to have 1-2 MT (the equivalent of 10 to 20, 90 kg sacks) or less corn as stored food for an entire calendar year. The corn is mostly stored as shelled kernels but may be left on the ear for a few weeks soon after harvesting to facilitate further drying. There are currently no officially designated sampling protocols for corn stored in the myriad of structures used in villages. With time and resources, probability-based sampling protocols could be developed. Nevertheless, sampling corn stored in village structures could be accomplished by adapting or extrapolating those aspects of the schemes discussed above that are applicable to the village foodstores.

Table 7.5 shows suggested sampling frequencies, composite sample sizes and laboratory sample sizes for different sizes of lots of corn. The lot sizes are expressed in terms of the common packaging sizes (50 and 90 kg sacks) and as estimated weight for corn on the ear. The incremental samples from each sack are collected from the bottom, middle and top when 10 or less sacks are sampled. For corn on the ear, an estimate of approximate weight is made and sufficient ears are collected to yield the indicated weight of shelled kernels (1-5 kg).

Table 7.5. Proposed[a] sampling frequency and sample size for small lots of corn in sacks

Weight MT	No. of Sacks	# of sacks to be sampled	Increments per sack[b]	Increments per lot	Increment size (kg)	Composite sample size (kg)	Laboratory sample size (kg)
0.09	1x90kg	1	5	5	0.2	1.0	1x1.0
–	1x50kg	1	5	5	0.2	1.0	1x1.0
0.45	5x90kg	5	3	15	0.2	2	1x2
–	5x50kg	5	3	15	0.2	2	1x2
0.900	10x90kg	10	3	30	0.2	6.0	1x5.0
	18x50kg	10	3	30	0.2	6.0	1x5.0
9.000	100x90kg	10	3	30	0.2	6.0	1x5.0
	180x50kg	13	3	39	0.2	7.8	1x5.0
90.00	1000x90kg	32	3	96	0.1	9.6	1x9.6
	1800x50kg	42	3	126	0.1	12.6	1x10.0

[a]Scheme designed to obtain sample weights within the mean weight range of last column in Table 7.4.
[b]Collected from top, middle and bottom of sack

With the question of *how much* to collect resolved, we are left with the more daunting task of *how* to collect the sample. Obviously, the ease with which a sample could be collected from a foodstore depends on the design of the storage structure. There are several forms of storage each with a slightly different level of accessibility difficulty. Storage techniques vary from hanging corn ears on a tree branch or placing the ears on a house rooftop to storage structures made from poles (Figure 7.1 B and C), sometimes held together by diamond wire (Figure 7.2A, upper right). Shelled corn may also be stored in completely sealed *Ferrumbu*-type mud cribs with a single release chute or port (Figure 7.1D).

Random sampling of some of the storage facilities in order to collect a representative sample is difficult as access is limited to the periphery, door area, or what can be released through the chute of a *Ferrumbu*-type crib. The degree of difficulty will vary from the relatively easily accessible rooftop or tree branch to the more difficult *Ferrumbu* crib. Much as it is acknowledged that emptying and refilling a mud crib would facilitate representative sampling, the mud-cribs are designed for one-time filling and multiple discharges as needed, through the chute. It is therefore, not logistically practical to collect several incremental samples from the *Ferrumbu* crib. The appropriate sample size would have to be collected through the chute in a single discharge. Alternatively, the household could be requested to let more grain flow out of the crib than required for the sample. In this way, a larger quantity of grain could be released, mixed, and a sample apportioned. The remainder is retained by the household.

For grain stored in sacks, collecting incremental samples poses its own challenges. Small diameter and easy-penetrating probes are not sufficiently long to traverse the length of the sack. The long and bigger diameter probes recommended by GIPSA, which are 1.5 to 3.7 meters long and 3.5 cm in diameter are extremely hard to push into a sack lying 10 or more rows below other sacks. Moreover, the most commonly available grain sampling probes in developing countries are ~ 0.4 m and ~1 m long. Furthermore, probe sampling is a slow process and the probes ultimately damage the sacks. If sacks should be probe-sampled when a lot is being moved, it is convenient to isolate those sacks selected for sampling, open the sewn end to permit easy penetration of the probe and returning the sacks to the consignment later.

Where practically possible, every effort should be made to collect several small incremental samples which, once combined, are sufficiently representative of the source. For village market samples (Figure 7.2B) use of an appropriate probe or hand grabs from the top middle and bottom of a sack should be adequate. Individual ears of corn collected from a designated portion of a farm can be considered as incremental samples as well.

Courtesy: Lauren Lewis

Figure 7.2. Corn cars on the ground and in a diamond wire storage crib (in rear) (A) and shelled corn in opened sacks at a village market (B).

Other Sampling Scenarios

Global transshipment of food commodities presents another complexity where appropriate equipment is not available. Bulk grain may be transported by ship, rail or truck (Figure 7.3). Grain in trucks and rail wagons can be sampled using appropriate length handheld gravity (open throat or compartmentalized) or pneumatic probes. Unlike trucks and rail wagons that have a relatively small capacity and shallow depth, a single cargo ship hatch can be 10 meters deep and hold 5,000 MT of grain. It is not generally advisable to probe a ship's hatch or bulk stored corn in sacks. Ships and stacks can therefore, only be appropriately sampled during loading (charging) or offloading (discharging), processes that might take several days. Test results from peripheral sampling of a stack of sacks of corn or from upper-layer probing of loose grain in a ship may be useful indicators of potential contamination: they cannot be used to draw inferences about the status of the entire stack or shipload.

Figure 7.3. Shelled corn in 50 kg sacks on a truck (A), on a stack in a warehouse (B) and as loose grain in a ship hatch (C), and 90 MT piles of fuzzy cottonseed (D).

In areas where cotton is grown, cottonseed is a bye product that is used as a feed ingredient, direct animal feed or as a source of oil. Fuzzy cottonseed is commonly stored in large, approximately 100 MT piles (Figure 7.3D). The standard sampling device for whole fuzzy cottonseed is a 1.2 m screw trier that draws 1.4 kg of seed per probing (14) yielding a composite sample of 14 kg. Pneumatic systems with longer probes (2.6 m) have been developed for sampling cottonseed (15,16) but are not yet widely used. The larger the cottonseed pile, the less likely that the analytical result obtained from the composite peripheral ten-probe sample reflects the true concentration of the entire pile. Similarly, the most appropriate time for sampling a cottonseed pile is when it is being made or pulled down.

Ready-to Eat Plate Foods and Snacks

Plate foods should be collected in their entirety, including any accompanying sauces and relishes, as a single sample in order to better estimate exposure. There is, however, an inherent problem with collecting plate foods from villages. It may require requesting the household to prepare double the quantity of food normally consumed to allow for collecting an adult-portion meal or to prepare a *sample-only* separate meal. It is debatable whether preparation of a *sample only* meal could bias the results assuming the household would prepare the specific meal from 'cleaner' corn. The likelihood is however, remote because the *sample-only* meal is likely to be prepared from the stock of corn meal used for preparing the household meal, unless the flour was completely depleted during the preparation of the family meal. To circumvent this potential bias, some investigators (17) have reported visiting sampling locations unannounced at meal times. How successful such an approach was, considering the social implications, is difficult to gauge.

SAMPLE PREPARATION AND HANDLING

Since the quantity of food held by a rural community household is usually small, reducing the collected composite sample to the required laboratory sample size may often not be necessary. More often, the investigator, aware that the smaller the number of items (increments) included in the sample, the higher the likelihood of an erroneous estimate of the aflatoxin content of the lot, may be inclined to collect more material as sample than is appropriate. However, there is no justification for collecting 25% of a household's food stock for the purposes of satisfying sample size.

When it is necessary to reduce the size of the composite sample, preference should be given to the option of initially grinding the entire larger composite sample. Grinding potentially breaks a single kernel into approximately a thousand or more tiny particles that distribute or spread more evenly within the entire lot than the larger whole kernels. Contamination originally occurring in one kernel would, after the grinding, occur in a thousand small particles that are easier to evenly mix into the larger lot. If grinding the entire composite sample is not practical, the kernels should be thoroughly mixed prior to the reduction process. Mixing corn kernels could be achieved using a spiked drum with a full radius lid on one side, a cement mixer where available, or any other device that may be deemed appropriate. The size of the drum would depend on the size of the sample. Metal or wood spikes are pushed to the center or across the entire diameter of the drum in several locations. The drum is half-filled with the composite sample, closed tightly and rolled back and forth several times over a distance of 5-10 times the drum circumference. The lid is removed and the corn kernels spread evenly on a flat surface in a circular form. Two perpendicular lines are drawn across creating four quadrants and kernels in opposite quadrants are combined into one portion. This process is called quartering (19). If the quantity is still large, the process of mixing and subdividing is repeated. The reduction scheme recommended above assumes lack of access to sample reduction devices such as the Cargo and Boerner dividers and related standard sample size-reduction equipment.

The efficiency of the mixing apparatus can be evaluated by coloring 10 to 12 corn kernels and adding them to the sample prior to the mixing event. Optionally, colored pebbles or beads of similar shape and size to the corn kernels could be used. The sample is run through the mixing process, spread evenly and divided into quadrants. The distribution of the colored corn kernels or pebbles among the quadrants is noted (numerical counted). If the colored kernels are not almost evenly distributed among the quadrants, the mixing time could be extended or the size or number of spikes changed. The efficiency of a cement mixer or any other device could be evaluated similarly.

Once a desired sample size is attained, the sample should be appropriately packaged to prevent deterioration (from moisture migration, insect infestation, further fungal infection, etc) during subsequent handling. Detailed labeling (sample number, estimated sample weight, common name of commodity, date of collection, collector's name, purpose, and source) is advised. Descriptions of the nature of the farm; stage of maturity; storage facility; form (i.e. ears, shelled whole kernels, grits, flour, etc); duration of storage; estimated total quantity of the lot sampled; and, if possible, weather conditions during the past week or month could be included. Furthermore, geographical coordinates using a global positioning system (GPS) would be useful. Ultimately, the type of information to be accrued and its intended use will determine the rigorousness of this process.

REFRENCES

(1)   FAO (1993) *Sampling Plans for Aflatoxin Analysis in Peanuts and Corn*. FAO Food and Nutrition Paper 55, Food and Agriculture Organization, Rome, Italy

(2)   Fonseca, H. 2003. *Brazilian J. Microbiol.* **33**: 97

(3)   EC (2006) *Commission Regulation (EC) No. 401/2006, OJ* **L70** *9.3.2006*

(4)   Hesseltine, C.W. (1983) in *Proceedings of the International Symposium on Mycotoxins*, September 6-8, 1981, Cairo, Egypt, National Research Center, Dokki, Cairo. 47

(5)   Whitaker, T.B. (2003) *Food Control* **14**: 233

(6)   USDA (2006) Inspecting Grain: Practical Procedures for Grain Handlers, U.S. Department of Agriculture, Washington DC.

(7)   Hongsuwong, T. (1989) in *Mycotoxin Prevention and Control in Foodgrains*, Semple, R.A., Frio, A.S., Hicks, P.A., & Lozare, J.V. (Eds), Food and Agriculture Organization, Rome, Italy

(8)   Lee, L.S., Lillehoj, E.B. and Kwolek, W.F. (1980) *Cereal Chem.* **57,** 340.

(9)   Cucullu, A.F., Lee, L.S., Mayne, R.Y., & Goldblatt, L.A. (1986) *J. Am. Oil Chem. Soc.* **43**, 89

(10)   Anonymous 2007. http://www.juran.com

(11)   Whitaker, T.B., Springer, J., Defize, P.R., deKoe, W. & Coker, R. (1995) *J. AOAC Int.* **78,** 1010

(12)   FDA 2007. Investigations Operations manual: Chapter 4 Sampling/Sample Schedules. U.S. Food and Drug Administration Office of Regulatory Affairs, Rockville, MD. http://www.fda.gov/ora/inspect_ref/iom/iomtc.html

(13)   Park, D.L. & Pohland, A.E. (1989) J. *J. Assoc. Off. Anal. Chem.* **72,** 399

(14)   Park, D.L. & Rua, S.M. Jr. (1991) *J. Assoc. Off. Anal. Chem.* **74,** 73

(15)   *Probe-A-Vac,* Tandem Products Inc., 520 Industrial Drive, Blooming Prairie, MN 55917

(16)   *Vac-A-Sample*, Seedburo Equipment Co. 1022 W. Jackson Blvd., Chicago, IL 60607.

(17)   Van Rensburg, S.J., Cook-Mozaffari, P., van Schalkwyk, D.J., van der Watt, J.J., Vincent, T.J. and Purchase, I.F. 1985. *Br. J. Cancer* **51**: 713

(18)   CX/FAC 2004. *CX/FAC 05/37/23. Proposed Draft Sampling for Aflatoxin Contamination in Almonds, Brazil Nuts, Hazelnuts and Pistachios. Annex 1.* Codex Alimentarius Commission, Joint FAO/WHO Food Standards Programme, Rome, Italy, 3.

(19)   Annonymous (2007) *Inspection and Sampling Procedures for Fine and Coarse Aggregate*s, Indiana Department of Transportation, IN, USA. http://www.in.gov/dot/pubs/q_assur/sample.html

8. QUALITY CONTROL MEASURES FOR MYCOTOXIN LABORATORIES

John Gilbert

Department for Environment,
Food and Rural Affairs
Central Science Laboratory
Sand Hutton, York
YO41 1LZ, U.K.

In recent years, one of the most significant developments in the analysis of foods from a perspective of food safety has been an increased emphasis on ensuring that data of demonstrable quality are being generated. This means that laboratories undertaking mycotoxin analysis should demonstrate that, when they report the presence of a mycotoxin, they have unequivocally confirmed its identity. Second, laboratories must show that they are quantitatively capable of obtaining an identical result for the same sample (± the reproducibility of the method) as a second independent laboratory. In practical terms, this means laboratories must have certain minimum and comparable analytical quality assurance (AQA) systems in place; they must use validated methods and demonstrate their analytical competence by taking part in proficiency testing. In Europe, these requirements are enshrined in legislation for official control laboratories (Additional Measures Food Control Directive) and, internationally, through Codex Alimentarius. In terms of the World Trade Organization (WTO), the concept of mutual recognition sets forth similar requirements that should be met by official laboratories. To resolve disputes involving foodstuffs for national or international trade, there is an obvious need for the results of testing to be comparable and accepted by both parties regardless of whether the analysis has been undertaken in the importing or the exporting country.

REQUIREMENTS OF AN AQA SCHEME

Logic and simple economics dictate that a measurement is worth making only if the result is meaningful (i.e., the result is fit-for-purpose). This premise applies regardless of whether one is dealing with a modern sophisticated high-technology laboratory or a less well-resourced laboratory in a developing country. In both cases, the relative cost of making any measurement is equivalent and, unless the result being generated is fit-for-purpose, such time and effort at best is wasted and at worst could be disproportionately damaging. Before making any analysis, the laboratory must define its requirements for the measurement; thus, if a mycotoxin test is being undertaken, the laboratory should know the purpose for which the result is being used. This will dictate whether a screening test is sufficient or whether it is necessary to use an official reference method to obtain an extremely accurate result close to the true value.

Whatever type of laboratory, and whether screening or very accurate analysis is being undertaken, an appropriate quality system should be in place. There are different levels of sophistication of quality control (QC) systems; for a laboratory in a developing country, a simple QC system can be introduced initially that can be used to form a basis for a future third-party-audited system fully in compliance with International Standards Organization (ISO) 17025. Good QC is fundamental to any laboratory and is as applicable to ensuring the quality of data generated by methods such as thin-layer chromatography as it is to results generated by sophisticated instrumental techniques such as liquid chromatography and gas chromatography/mass spectrometry. The basic principle of AQA involves a simple common-sense approach. It often involves practices that already exist in good laboratories but have not always been well documented. Thus, AQA is about the following:

- Having suitably trained staff undertaking the analytical work

- Having documentation (training records) to demonstrate compliance

- Documenting the method of analysis and its performance (preferably by a validated method)

- Documenting all control measures that are to be applied

- Traceability of measurement: showing how the calibration was obtained and how the calibrant can be traced back to some absolute reference point

and

- Using suitable instruments and documenting their maintenance and calibration.

All these measures can be audited by a third party, which is really what constitutes formal accreditation, or they can be in place but checked only internally. Finally, AQA is about demonstrating that your laboratory can get the "right result" when you analyze "blind samples." Although this can be done with internal quality assurance (QA) samples, the most satisfactory way is through participation in an external proficiency testing scheme.

ESTABLISHING A QA SYSTEM

The first step in establishing any QA in a laboratory is to decide which system to use. The choice ranges from an informal self-monitored system to a formal third-party-audited ISO 17025 accreditation to Good Laboratory Practices (GLP) (1). Broadly, accreditation is best suited to laboratories that undertake measurement (i.e., analytical laboratories), as it is method based, whereas GLP is more suited to laboratories that undertake toxicological testing, as it has a stronger emphasis on documentation and record keeping as opposed to measurement traceability. Whichever choice is made, the laboratory needs to have management committed to establishing a QA system, and the decision needs to take cognizance of the existing experience as well as the ultimate suitability and fitness-for-purpose of the system that is introduced.

Accreditation is usually based on specific methods—for example, determination of aflatoxin by liquid chromatography. If a laboratory is going to expend time and resources seeking accreditation for a specific method, it needs to be one that is well established and is used on a routine basis. It is possible to get generic accreditation (e.g., for liquid chromatography/mass spectrometry analysis), which is useful for service functions and for between-team working; it has the additional advantage that it can be used in a research function. It also needs to be decided which methods to accredit. This depends on the nature of the laboratory and for whom the laboratory undertakes work. In general, accreditation is important when dealing with third parties, whether customers or other laboratories, with whom a dispute may arise about comparability of results. For commercial testing laboratories, and within the European Union, there may be no business without accreditation—i.e., formal requirements in some markets are such that some laboratories cannot function in a commercial environment without appropriate accreditation in place.

DOCUMENTATION REQUIRED FOR ISO 17025 ACCREDITATION

There are basically three levels of documentation required for accreditation:

Quality Manual for the Laboratory

This is the master document for which only one is required, regardless of the size of an organization. Thus, an organization may have 20 individual laboratories, each requiring accreditation in different areas of chemistry and microbiology, but a single quality manual suffices for all these laboratories, if covered under the same accreditation. Apart from giving the organization structure, the quality manual clearly sets out things such as quality policy, organization, and management as well as the processes the organization has in place for undertaking quality audits and quality system reviews. Areas such as handling complaints are covered in the quality manual.

Technical Manual for Sections of the Laboratory

For each discrete section that has a distinct management structure, it is necessary to have a technical manual. For different sections, these manuals may in fact have many common elements and a common format should be adopted as well as a consistent approach. The technical manual covers topics such as:

- Equipment,

- Calibration and measurement traceability,

- Test methods and procedures,

- Laboratory accommodation and environment,

- Handling of test items,

- Records, and

- Test reports.

Standard Operating Procedures (SOPs)

Each individual analytical method (e.g., an analytical procedure for ochratoxin A in coffee) is described in the form of an SOP. It is not sensible to be prescriptive in terms of what should be in an SOP, but normally the following elements are found:

- Scope: what analytes it covers and what matrices (any limitations),

- Safety aspects,

- Principle of method,

- Sampling/sample preparation,

- Reagents,

- Apparatus,

- Method (extraction and cleanup),

- Determinative step (chromatography and quantification),

- QC procedures, and

- Appendices: specific operating instructions.

ESTABLISHMENT OF QUALITY INFRASTRUCTURE

The basic infrastructure for establishing and maintaining accreditation depends on a series of checks and cross checks, which are carried out at regular intervals both internally and by third parties. From the time all documentation is complete and the laboratory is effectively running within an accredited environment, the following sequence of events needs to occur to establish formal accreditation:

- Internal audits,

- Application to accreditation body and payment of fee,

- Inspection of facility and records by accreditation body,

- Pass/fail inspection,

- Correction of noncompliances,

- Accreditation granted for specified scope: number issued and permission to use logo,

- Continuing internal audits,

- Accreditation body carries out surveillance visit: about 12 months after initial inspection, and

- Possible extension of scope.

Noncompliances

During an accreditation assessment visit, noncompliances may be found that fall into a number of different categories depending on the seriousness of the shortcoming. Assuming that a major failure is not found, these noncompliances can be overcome by taking remedial action. However, there needs to be evidence of the action taken to meet the assessor's requirements that may take the form of:

- Additional data,

- Changes to documents, or

- External calibrations.

Equipment Requirements for ISO 17025

The laboratory shall be furnished with all equipment items required for correct performance of the tests and/or calibrations (including sampling, preparation of test and/or calibration items, processing and analysis of test and/or calibration data). Equipment used for testing must be capable of achieving the required accuracy. Equipment significant to the tests performed should be uniquely identified and records must be maintained that include at least the identity of the equipment by manufacturer's name, type identification, and serial number. The equipment must comply with specifications, and documentation should identify its current location. All equipment must be verified/calibrated on receipt and after any move. The manufacturers' instructions should be available (or their location noted). As part of the accreditation, equipment should be operated only by authorized personnel. Instructions on the use and maintenance (manuals and SOPs) must be available.

Calibration programs shall be established where there might be a significant effect on the result. Records should be kept of dates, results and copies of reports of calibrations, adjustments, acceptance criteria, and due date for next calibration. There need to be a maintenance plan and records of any damage, malfunction, modification, or repair to any equipment maintained. The maintenance and repair records can be in the form of instrument logbooks. The logbooks, whether for single items (balance) or for an instrumental setup (liquid chromatography pump, autosampler, detector, etc.), should contain information on identity, details of damage, repairs, description of defects, component changes, and performance checks and results. Logbooks should also contain procedures for calibration; acceptance limits of results; references to operating instructions; results of performance checks; indication of acceptability; and records of use, date, time, and details of the operator. For other items of equipment, more specific conditions may apply.

General Housekeeping

- General equipment should be kept clean and suitably maintained with periodic performance checks if it can be adjusted and the adjustment can influence the result (e.g., water baths).

- The quality of volumetric equipment such as glassware should be specified. It should be checked before use and kept clean and damage free. Instructions on storage and cleaning should be stipulated.

- The performance of measuring equipment must be regularly checked as correct operation, maintenance, etc. are not a guarantee of correct functioning. It is necessary to decide on the frequency of checks, which can be decreased when stability has been demonstrated.

- Reference standards should be used for calibration purposes only and should be protected and stored separately from similar equipment. There must be traceability, and current certificates should be available.

- Computers' performance should be ensured by using the same principles as for other equipment; again, records, software installation, version, updates, validation of software, continuous control (passwords, templates), and safekeeping of data should be demonstrated and documented.

- Equipment failing to meet acceptance criteria and broken/ damaged equipment must be withdrawn from use. Such equipment should be marked clearly and removed from the area. Such equipment should be fully tested and calibrated before it is reintroduced for use.

Personnel Requirements for ISO 17025

Laboratory management shall ensure the competency of all who operate specific equipment, perform tests, evaluate results, and sign test reports and calibration certificates. When staff are undergoing training, proper supervision shall be provided. Personnel performing specific tasks shall be qualified on the basis of appropriate education, training, experience, and demonstrated skills. The laboratory shall use personnel who are permanently employed by or under contract to the laboratory. Contracted personnel must be competent and supervised and must work in accordance with the quality system. The management shall formulate the goals with respect to the education and skills of the laboratory personnel. The laboratory shall have policies and procedures for identifying training needs and providing training of personnel. The training program shall be relevant to the present and anticipated tasks of the laboratory. The laboratory shall maintain current job descriptions for managerial, technical, and key support personnel involved in tests.

Responsibilities may be defined in the following areas:

- Performing tests/calibrations,

- Planning tests/evaluating results,

- Method modification/development,

- Validation of new methods,

- Expertise and experience,

- Qualifications and training programs, and

- Managerial duties.

Management shall authorize specific personnel to perform particular types of sampling and tests, issue test reports, give opinions and interpretations, and operate particular types of equipment.

Maintaining Staff Records

Staff records shall be maintained with respect to the following aspects: relevant competence, educational and professional qualifications, training skills and experience, date when authorization/competence was confirmed, and criteria on which it is based.

Technical Manager Responsibilities

The technical manager is responsible for ensuring that only suitable, documented, and appropriate methods are used. The technical manager is responsible for documenting the operating instructions for the available equipment. The technical manager must ensure that all laboratory assistants and technicians clearly understand and master analytical work. The technical manager is responsible for regularly monitoring the quality of the work undertaken using suitable QC measures.

The technical manager must also ensure the following:

- Results are documented and archived,

- Results are reported according to established rules,

- Staff receive training in quality assurance,

- He/she is completely familiar and follows laboratory quality policy, and

- He/she follows developments within his/her analytical field.

In addition to the technical manager, all other laboratory staff have responsibilities. They must assume the following responsibilities:

- Closely follow given instructions and analytical methods,

- Document all analytical work,

- Be acquainted with and follow security and environmental regulations,

- Be acquainted with and follow the quality policy of the laboratory.

Calibration Procedures

Accreditation requires that a calibration program is in place. Reference standards of measurement must be calibrated by a body that can provide traceability to a national standard. Verification is required between annual calibrations. The calibration status should be marked on all equipment. The frequency of calibration checks depends on the nature of the particular instrument:

- Annual checks: e.g., for reference thermometers,

- Six-month checks: e.g., for external calibration of balances and pipettes,

- Monthly checks: e.g., for pipettes,

- Weekly checks: e.g., for refrigerators and freezers and

- Daily checks: e.g., for balances, pH meters, etc.

Balances can be calibrated internally by appointed staff with proper training. Weights used must be calibrated and traceable to a national or international standard, and, although the calibration is carried out annually, performance must be checked at least weekly with standard weights. Standard weights should be checked at the time of the annual calibration. For pipettes, the frequency of calibration depends on the extent of use. Acceptance limits for measurement equipment must be set. The limits must take into account the accuracy requirements of the method. For thermometers that are glass-liquid stable, calibration is necessary only every 5 years, although they must be checked annually. Electronic thermometers must be calibrated annually and checked quarterly. A reference thermometer is used only for calibration purposes, traceable to an international standard. External calibration is carried out by an accredited agency.

INTERNAL QC

Internal QC (IQC) is an essential aspect of ensuring that data released are "fit for purpose." IQC also enables monitoring of the quality of data on a run-by-run basis, and any runs that are outside acceptable limits should be rejected. The basic approach in IQC involves analyzing control materials alongside test materials. The outcome of such analyses forms the basis for any decisions about the acceptability of test data. To gain maximum benefit from IQC, it is important to have a good appreciation of the nature of errors, which can be random or systematic. Random errors are random positive and negative deviations of results from the true values, whereas systematic errors typify a persistent bias that may be identifiable only over a long period of time. Systematic errors may be tolerated if they can be kept within prescribed bounds. Other terms that are frequently used are run effect (deviation of the analytical system during a particular run) and within-run precision (a term that refers to variance occurring during an analytical run).

In fact one has only a limited indication of within-run precision based on duplication within a run of measurements, even if duplicate samples are placed at random in a run. Duplicates should be complete (blind) independent analyses of separate test portions. Duplicate injections of the sample extract is ineffective, as it does not control the rest of the method.

Control charts are an essential element of IQC. Control charts are obtained when values of concentration measured on a control material are plotted on a vertical axis against run number on the horizontal axis. For control purposes, horizontal lines are included to represent the following: μ, mean or assigned value; $\mu \pm 2$ standard deviations, warning limits; $\mu \pm 3$ standard deviations, action limits.

Standard deviations should be estimated with care. They should include a between-run element. Estimates based on a single run would produce unduly narrow control limits. For a system in statistical control, 1 in 20 values will fall outside the $\mu \pm 2$ line (i.e., the warning limit), 1 in 1000 values will fall outside the $\mu \pm 3$ line (i.e., the action limit). Control charts can be used to assess analytical trends, which is particularly important with ISO 17025.

CONTROL MATERIALS

Control materials must contain an appropriate concentration of the analyte and the concentration must be assigned. The materials should be as representative of the sample as possible—i.e., the same matrix in the same physical form (e.g., same state of comminution). Control materials must be adequately stable over the period of interest, and it must be possible to divide material into effectively identical portions for analysis. Large amounts may be required to use over an extended period of time.

Certified Reference Materials

Certified reference materials (CRMs) are the ideal control materials because they are completely traceable to an independent source. However, frequently the correct matrix/analyte may not be available or, if available, the cost may be prohibitive. The CRM may not be applicable if the matrix or analyte is unstable, or the CRM may not be available in sufficient quantities or for an extended period.

Two powdered milk reference materials are available from the European Community Bureau of Reference Materials (BCR), which have certified low-level contents of aflatoxin M_1. Peanut butters with certified levels of individual aflatoxins are also available from BCR, as are wheat and maize samples naturally contaminated with deoxynivalenol, animal feedstuffs containing certified levels of aflatoxins, and wheat containing ochratoxin A. Projects are under way to develop cereals with certified contents of fumonisins and zearalenone. This increasing range of certified mycotoxin reference materials should offer greater possibilities for rapid assessment of new methodology, for different toxins and different matrices, in the future. However, in those instances when a CRM is not available, it may be necessary to prepare in-house reference material.

Materials Validated in Proficiency Testing

These materials have been analyzed by many laboratories; thus, the consensus mean can be regarded as a validated assigned value with a meaningful uncertainty. However, one must be aware of the method(s) that may have been used in the proficiency testing in case of bias. Participating laboratories should not retain surplus proficiency test materials because, strictly, they do not have a traceable value as a result of probable instability of the analyte. It is unlikely that stability would have been established.

In-House AQA Material

For any naturally contaminated sample, laboratories can prepare their own large batch of material and make an assigned value by careful analysis. The laboratory must take care to avoid bias, and there is an obvious need for an appropriate analytical method. Independent validation may be a good idea, and this can be obtained by analysis by another laboratory or by another method, ideally one based on a different principle.

Spiked Control Materials

Spiked control materials have the advantage that the matrix is identical to that of the test materials and that the concentration of analyte is known. The disadvantage of spiking is that any interaction with the matrix, such as binding, which may occur with naturally contaminated samples, may be absent in spiked controls. Spiked control samples are particularly useful for making recovery checks, and they are useful when the analytes or matrices cannot be stabilized. Poor recovery with spiked samples is a very strong indicator that the method will perform poorly for naturally contaminated samples.

Blank Determinations

Blank determinations are an essential part of the analytical process. The reagent blank is the simplest. It is the process by which one carries out the full analytical procedure except for adding the test portion. With the procedure blank determination, a sample known to be free of the analyte (or with analyte content below the detection limit) is analyzed. Inconsistent results in either of the above instances might suggest sporadic contamination of the equipment; hence, the results of a true analysis may be questionable.

PROFICIENCY TESTING

Proficiency testing is a continuous objective assessment of the ability of a laboratory to produce accurate and reliable results. In proficiency testing, laboratories receive samples for analysis at regular intervals, report the results to the scheme organizers, and then receive an assessment of their performance. The identities of the laboratories involved are kept confidential, although each laboratory receives a report with useful indications about the overall performance of all participants and information on performance related to the methods of analysis that have been used.

Proficiency testing should be regarded as an integral part of accreditation and should be seen as providing valuable information for internal purposes as well as a laboratory performance indicator to third parties. In proficiency testing (unlike collaborative trials), all participants use the methodology with which they are accustomed and which is in everyday use in their laboratory. Ideally, both the matrix and the analyte should be something that is routinely tested in the laboratory. Participants should be discouraged from undertaking proficiency testing with unfamiliar materials. Frequently, however, the desired matrix may not be available and the best approximation has to suffice.

The requirements for establishing and running proficiency testing schemes are stipulated in an ISO/IUPAC/AOAC International

Harmonized Protocol. Nationally and internationally, a number of commercial schemes are run on a regular and systematic basis. For example, in the U.K., a proficiency testing scheme called FAPAS (food analysis performance assessment scheme) has been operating since 1990 and has included mycotoxins among the analytes that are tested. FAPAS has expanded rapidly since inception and, as of mid-1998, some 600 participant laboratories from 50 countries were taking part. The World Health Organization Global Environment Monitoring Scheme (GEMS) program has also organized proficiency testing for mycotoxins, but sample distribution has been more sporadic.

Proficiency testing schemes require procuring or producing materials that closely resemble those encountered in practice. The most suitable is material frequently analyzed by participating laboratories. After blending or preparation into a suitable form for distribution, homogeneity testing needs to be carried out to ensure uniformity of samples. Test materials are distributed to participants, and analysis is undertaken by the normal method of analysis; within 1 month, results are reported back to a secretariat. The results must be processed rapidly for maximum benefit to participants. In FAPAS, a report is issued within 1 month giving all results for the round plus individual performance scores (known as z scores) for each analyte. The z scores are calculated based on the "true" value of the analyte and the standard deviation expected at that concentration (from collaborative trial data or the Horwitz curve). A z score of +2 is deemed satisfactory, a z score between –2 and –3 or +2 and +3 is deemed questionable, and outside that range is unsatisfactory.

REFERENCE

(1)   *Monograph 45*, Organization for Economic Cooperation and Development, Paris, France

FURTHER READING

(2)   Key, P.E., Patey, A.L., Rowling, S., Wilbourne, A., Worner, F.M. (1997) *J. AOAC Int.* **80**, 895

(3)   Sargent, M., & MacKay, G. (Eds) (1995) *Guidelines for Achieving Quality in Trace Analysis*, Royal Society of Chemistry, Cambridge, U.K.

(4)   Thompson, M., & Wood, R. (Eds) (1993) *Pure Appl. Chem.* **65**, 2123

(5)   Thompson, M., & Wood, R. (Eds) (1995) *Pure Appl. Chem.* **67**, 649

(6)   Wood, R., Nilsson, A. & Wallin, H. (1998) *Quality in the Food Analysis Laboratory*. RSC Food Analysis Monographs, Royal Society of Chemistry, Cambridge, U.K.

9.   METHOD VALIDATION PROCEDURES

Albert E. Pohland

AOAC INTERNATIONAL
481 North Frederick Avenue, Suite 500,
Gaithersburg, Maryland 20877

It is generally believed that, to analyze a sample of food or feed with some confidence in the obtained result, one must have access to: (a) a sample representative of the bulk lot; (b) a validated method; (c) an analyte reference standard; (d) a qualified analyst; and (e) a laboratory under quality control, preferably one that has been accredited following International Standards Organization (ISO) Guide 17025 (1). All these requirements are important; however, the subject of this chapter is method validation.

What is meant by method validation? The most frequent response to this question is that it is the process by which the capabilities of a method are assessed. This is normally done originally by the developer of the method; however, users of the method also have the responsibility of validating the method to ensure that it meets their particular requirements and that the analyst has the experience and qualifications to perform the method competently.

According to ISO (2), validation involves "confirmation by examination and provision of objective evidence that the particular requirements for a specified intended use are fulfilled." This concept that method validation is the set of tests used to show that a particular method is "fit for purpose"—i.e., suitable for its intended use—is clearly enunciated in many of the documents on method validation (3, 4). Clearly then, the degree of validation required depends on the intended use of the method. Table 9.1 provides examples of frequently encountered uses for analytical methods, arranged according to the degree of validation frequently observed in practice. Be aware that methods used for regulatory purposes are frequently found in all three categories.

Table 9.1. Relationship between degree of validation and intended use

Single Laboratory validation	• For research purposes • To be used only once on a limited number of samples or in a limited number of laboratories • To be used by a single laboratory–e.g., a quality control or calibration laboratory • To ensure reliability of a method imported from another laboratory • To ensure proper use of a previously validated, off-the-shelf method • Use in situations where there is a lack of participating laboratories for collaborating study • Multiresidue methods where collaborative study is impractical
Multilaboratory validation (two to eight laboratories)	• Use where only a few laboratories have a need • Use where full collaborative study is not justified for economic reasons • Use where full collaborative study is not justified because of an immediate need for quick validation • Use for relagatory surveillance activities
Multilaboratory validation (more than eight laboratories)	• Use as "official methods" (AOAC) or "reference methods" • (Codex: Category2) • Use in dispute situations • Use as the "gold standard" against which other methods are compared

To satisfy oneself that a method is fit for its intended use, it is necessary to examine the method's performance parameters and make an estimate of the method's expected measurement uncertainty. The performance parameters of a method normally include attributes such as applicability, recovery, detection/determination limits, operating range, false-positive and false-negative rates, ruggedness, accuracy, confirmation of analyte identity, selectivity, sensitivity, specificity, reproducibility, and repeatability. The procedures for calculating measurement uncertainty are many and variable at the moment. The reader is referred to ISO (5), Eurachem

(6), and AOAC (7, 8) guides for assistance. It must be appreciated in addressing the issue of measurement uncertainty that one is attempting to estimate the confidence one might have in an experimentally determined analytical result. There is also the confidence one might have in the use of the method and the measurement uncertainty of the method (as distinct from the measurement uncertainty of a specific analytical result), which is usually determined by conducting an interlaboratory collaborative study (ring test). It is incumbent on the laboratory using a new method not developed in that laboratory to ensure that it can use the method within the parameters and measurement uncertainty reported by the developers of the method. This is frequently done through the analysis of check samples or certified reference materials or by comparing analytical results obtained using the method with those obtained using a reference, or "official," method.

In this chapter the typical processes and requirements for validating methods in all three categories are described.

CATEGORY I: SINGLE LABORATORY VALIDATION

There are many practical situations in which single laboratory validation is sufficient (Table 9.1). In determining whether a particular method is fit for use, the analyst must evaluate certain performance characteristics of the method through laboratory study. This is true with a newly developed method as well as with a method imported from another laboratory. It is incumbent on analysts to ensure that the methods in their hands— that is, when properly performed by them—give results that meet the requirements of the intended use. Therefore, method validation is an integral part of a laboratory's quality assurance program.

If the method is not new, the degree of validation—that is, the number of performance parameters selected for evaluation—will depend on the extent to which the method has been validated previously. Methods that have been previously validated by a consensus standard setting organization, such as AOAC INTERNATIONAL or ISO, where full (9) collaborative studies are used, need not be validated to the same extent as new methods, where complete validation is necessary; in these cases, a limited number of performance characteristics are sufficient to ensure the analyst that the method is "fit for use" (10). It is important to note that all laboratory results used to validate a method must be properly documented.

In single laboratory validation schemes, the following performance characteristics are generally included, although not always in the order listed.

Applicability

This information is usually contained in the scope statement accompanying the method and should include a clear summary about the analyte/matrix combinations for which the method was developed as well as the applicable concentration range. In some cases, it is important to address the issue of speciation—e.g., total mercury or methyl mercury, bound or free cadmium, etc. If the method is to be applied to analyte/matrix combinations not identified in the scope, or if the method has been modified, then additional data are needed to ensure that the performance characteristics reported by the author of the method still pertain.

Ruggedness

Ruggedness testing is an integral part of method development. Its purpose is to identify those elements of the method that are particularly sensitive to relatively minor changes in technique, reagents, and environmental factors. The usual procedure (10, 12) is to introduce those changes and observe the effect on the accuracy and precision of the analytical results. Any serious deviations in accuracy and precision indicate a critical control point in the method; these factors then need to be carefully controlled by the user of the method to obtain acceptable results. Once ruggedness testing is done, and the critical control points are identified, it is rarely repeated in subsequent validation-for-use exercises unless the method has been significantly modified.

Selectivity/Specificity

These two terms are frequently used interchangeably; they refer to the confidence the analyst has in the method's capability to determine a specific analyte. In the development of the method, the analyst must show that the signal or end point used is due only to the analyte to be determined and not to some interference. In the case of chemical analyses, such interferences may be compounds of similar chemical structure or chromatographic characteristics. Usually the developer of the method includes a confirmation of identity step, such as a second chromatographic procedure (e.g., use of different polarity columns or thin-layer adsorbents), or confirmation by mass spectrometry or infrared spectrophotometry (10). It must be acknowledged that it is impossible to prove the absence of interfering substances, but it is incumbent on the analyst to do his best to ensure that the most likely interferents are not present by analyzing test portions to which such substances have been added.

Analytical Range

Most analytical methods are developed and validated for a specific use over a defined concentration range, and this is stated in the scope of the method. This range frequently is stated in terms of a limit of quantitation, a limit of determination up to some upper bound of applicability. Within this concentration range is usually an area in which the plot of analyte response (i.e., the property measured in the method) vs. the analyte concentration is linear; this is referred to as the working range and is usually obtained by conducting recovery studies on spiked materials.

In validation studies with the harmonized guidelines for single laboratory validation (12), the following are recommended: (a) that six or more concentration levels be used; (b) that the levels be spread evenly over the concentration range expected to be encountered—e.g., if the method is to be used to meet some regulatory limit, then concentrations on both sides of the limit should be selected; and (c) that the tests be run in duplicate in random order. Regression analyses are frequently used to compare the linearity of the data with that reported by the originators of the method.

Studies of the analytical range frequently include evaluation of matrix effects. This is usually done by the method of standard addition to the extract from the test material, in which the analyte is added at the same concentration levels as that reported in the method calibration. If the two calibration curves are identical, no matrix effect is apparent.

Limit of Detection

There is no generally agreed definition for detection limit. In general, it involves the lowest concentration of analyte in the test material that can be unambiguously measured by the method. It is frequently associated with the "instrumental detection limit"; however, this limit is invariably far lower than that which can be determined in the analytical sample. In practice, the detection limit is usually determined by "repeat analysis of a blank matrix, and is the analyte concentration whose result is equivalent to 3 times the standard deviation" (3). The harmonized protocol for single laboratory validation (12) recommends six independent complete determinations in making this estimate. It must be appreciated that, unless an inordinate amount of effort is exerted, the variability associated with analyses at the detection limit will be extremely large, and, consequently, reporting of such results will be open to challenge.

Limit of Determination (LOD)

Perhaps a more useful performance characteristic of a method is the limit of determination; although, again, there is no universally accepted definition for this element. According to the harmonized protocol for single laboratory validation (12), it represents "the concentration below which the analytical method cannot operate with an acceptable precision" [e.g., relative standard deviation (RSD) = 10% or sometimes 2 times the detection limit]. Garfield *et al.*, (8) define it as "the lowest concentration of an analyte, using an appropriate standard or sample, that can be determined with an acceptable level of accuracy and precision." The use of determination limits in method validation is not recommended by the harmonized protocol.

Limit of Quantitation (LOQ)

This is similar to the LOD and is the "minimum concentration of the analyte in the test sample that can be determined with acceptable precision (repeatability) and accuracy under the stated conditions of the test" (10). Usually, the LOQ is determined with recovery data and frequently represents the lowest concentration level of the linear range determined experimentally.

Accuracy

This term is used synonymously with "trueness" and "bias." It is a measure of the closeness of agreement between an analytical result and the accepted reference value. Sometimes, distinction is made between the types of bias encountered in the laboratory— method bias, run bias, and laboratory bias. To be clear about what is being reported or discussed, I suggest the following: (a) that bias be defined as the difference between a single result from the true value, (b) that trueness represent the difference between the average of a series of determinations and the true value, and (c) that method accuracy be the difference between a long-term average and the true value. Practically, the total bias is the issue and is referred to as accuracy.

The best approach to measuring accuracy is to use certified reference materials (CRM). Certified reference materials are generally traceable to an international standard whose concentration is known and accepted as the reference value with a known measurement uncertainty. In practice, a series of blanks and samples (the Eurachem guide recommends 10 each) are analyzed and the mean of the results is compared with the reference value.

More often than not no CRM is available, and the laboratory must prepare the test sample by spiking the matrix with a CRM or with a well-characterized analyte of suitable purity and stability. In some cases, a proficiency test sample is a useful alternative.

A common, frequently used procedure for measuring accuracy is to spike a sample with a known amount of analyte, analyze the sample, and calculate recovery of the spike. Recoveries close to 100% indicate a lack of bias. However, there is no guarantee that a naturally contaminated matrix would behave similarly; therefore, 100% recovery does not guarantee accuracy, but poor recovery certainly indicates a lack of accuracy.

Precision

Precision is a measure of the closeness of agreement between independent test results—i.e., between two or more complete analyses. The most commonly reported measures of precision are the data obtained from duplicate analyses of the same material (although both samples are usually presented to the analyst as unknowns). This type of precision is often referred to as repeatability and is reported as a standard deviation (s) or relative standard deviation within laboratory (RSD_r). The methods for calculating these functions are well known (8, pp. 29–46). Because precision is known to vary with analyte concentration, it is important to check precision at both the high and low end of the linear range of the method. Repeatability is usually equivalent to the smallest measure of precision, indicating the closeness of agreement between the analytical results that can be expected by a single analyst using a single piece of equipment, usually over a short period of time, within a single laboratory.

Of course, the more analysts, the more analyses, and the more laboratories that analyze the sample, the more meaningful the precision estimate. When more than one laboratory is involved, the precision is referred to as reproducibility (RSD_R); the relative standard deviation between laboratories (RSD_R) is usually 2–3 times the repeatability (RSD_r).

It is important to note that precision statements calculated in this way can be obtained only in quantitative analyses. When a qualitative analysis is encountered, basically a yes/no situation, a measure of precision is found by calculating the false-positive and false-negative rates.

Determining the performance characteristics of a method is only the first step in method validation; however, it is often the most expensive, time-consuming, and laboratory-intensive part of the validation exercise. See Fajgelj and Ambrus for typical examples of single laboratory method validation for pesticide methods (10, pp.188–203) and for veterinary drug methods (10, pp. 204-217). The second step in method validation is an independent peer review of the data; this is usually done in single laboratory method validation by the laboratory director or quality assurance manager.

CATEGORY II: MULTIPLE LABORATORY METHOD VALIDATION (TWO TO SEVEN LABORATORIES)

All the key performance parameters required for single laboratory validation (see above) are also required for this category of validation. The only difference is that samples are exchanged between laboratories. Normally, the larger the number of laboratories involved the better (narrower) the precision estimates. There are many examples of multilaboratory validation programs involving collaborative studies (ring tests). The AOAC Peer Verified Methods (PVM) Program is a typical example (13). A characteristic of all such validation schemes is that more than one laboratory is involved. It is important that the analyses be independent and preferably blind—i.e., the analyst does not know the sample concentration or that the sample is part of a collaborative study. The sample is treated like any other sample analyzed by the laboratory. The data generated in these tests allow one to calculate an interlaboratory relative standard deviation (RSD_R)—i.e., method reproducibility.

For an example of the procedures followed in the validation of a multilaboratory collaborative study, see Table 9.2, which outlines the steps followed in the AOAC's PVM validation program. In these studies, the results are reported to the study director and peer reviewed by the general referee with the assistance of two outside, independent reviewers.

Table 9.2. AOAC PVM Program (13)

Step 1: Determine scope Purpose Type Applicability	Enforcement, surveillance, etc. Reference, screening, surviellance Analyte/matrix, concentration range
Step 2: Ruggedness testing	Identify critical control points
Step 3: Determine performance characteristics	Selectivity/specificity Analytical range: calibration curve, inearity Limit of detection/quantitation Accuracy: recovery Precision: repeatability/reproducibility
Step 4: Prepare method description	Follow PVM format
Step 5: Develop protocol for testing in second laboratory	See checklist for development of independent laboratory protocol
Step 6: Select second laboratory	See guidelines for selection of second laboratory
Step 7: Second laboratory tests method	Second laboratory evaluates performance characteristics
Step 8: Second laboratory submits report to originating laboratory	
Step 9: Originating laboratory prepares summary report and submits to AOAC for evaluation	AOAC asks technical referee to evaluate; technical referee selects two experts to review the data. Based on these reports, the method is accepted.
Step 10: Technical reviewer informs AOAC of decision	AOAC publishes in PVM Notebook Review every 5 years

CATEGORY III: MULTIPLE LABORATORY VALIDATION (EIGHT OR MORE LABORATORIES)

Some years ago the AOAC realized there was a need, statistically, for a minimum amount of valid data to confidently calculate the interlaboratory relative standard deviation (RSDR). This need developed into the AOAC's Official Methods Program, which has been used in the United States to identify official methods for regulatory use.

This standard has been accepted internationally and is now used frequently to identify reference methods (e.g., Codex Category II methods). Internationally accepted guidelines have been published for the design, conduct, and interpretation of what are often referred to as full collaborative studies (9). According to these guidelines: (a) a minimum of five materials are required (for a single level specification and a single matrix, this may be reduced to three), (b) a minimum of eight laboratories reporting valid data are required (in special cases involving expensive instrumentation or specialized laboratories, this may be reduced to five), and (c) a minimum of two replicates are required (if repeatability is desired). For qualitative analyses, the AOAC has some additional requirements (not part of the harmonized protocol): a minimum of 10 laboratories, two analyte/matrix combinations, five test samples per level, and five negative controls per matrix.

A general outline of the steps involved in conducting a collaborative study following the AOAC's Official Methods Program is presented in Table 9.3. Conducting a collaborative study is expensive, resource intensive, and time-consuming. Therefore, it is important to observe some basic advice: (a) optimize the method and perform ruggedness testing before running the study; (b) take care to clearly and understandably prepare the method description; (c) develop a protocol and obtain approval of a statistician (or the AOAC) before beginning the collaborative study; (d) give careful attention to the test samples distributed to the collaborating laboratories—homogeneity, analyte stability, random number identification, etc.; (e) carefully select the participating laboratories to ensure the analysis are performed by qualified analysts, working in a good laboratory quality control environment.

Table 9.3. AOAC Official Methods Program (9)

Step 1: Preliminary work (within one laboratory)	Type of method, probable use, analyte/matrix
Determine purpose and scope	Conduct trials changing one variable at a time
Optimize method	Ruggedness testing
	Selectivity/specificity
Develop within-laboratory performance	Analytical range, calibration surve, linear range
Characteristics	Limit of detection/quantitation
	Accuracy: recovery
	Precision: repeatability
Step 2: Prepare description of method	
Step 3: Conduct collaborative study	Prepare study protocol; submit for AOAC approval
	Identify/select collaborating laboratories
	Prepare and send formal letters of invitation to join the study
	Prepare instructions and report forms
	Prepare test samples, blanks and practice samples; observe proper storage and handling
	Provide for reference standards
	Set time limits for completion of study
	Send samples to collaborators
Step 4: Collate/review data from collaborators	Analyze data statistically
	Prepare report in publication format
	Calculate recovery, RSD_R, and HORRAT
Step 5: Submit report to all participants for comment	
Step 6: Submit report to AOAC for peer review	Review general referee, methods committee, statistics/safety committees, official methods board

Acceptance Criteria

Once the performance parameters of a method are known, the question becomes one of method acceptability. Acceptance requirements vary depending on the validation scheme followed and, in part, the intended use to be made of the method. The AOAC Official Methods Program tends to use the acceptance criteria presented in Table 9.4. Acceptance criteria proposed for single laboratory validation of pesticide residue methods and methods for veterinary drugs are shown in Table 9.5. They are summarized here only as a guide, because there is currently no internationally harmonized agreement relative to acceptance criteria.

Table 9.4. Suggested acceptance criteria for official methods–AOAC (14)

		Chemistry		
Concentration	Recovery (%)	RSD_r(%)	RSD_R(%)	HORRA T*
1 ppb (1 ng/g)	TBD	TBD	TBD	0.5-2.0
10 ppb (10 ng/g)	60-120	20	30	0.5-2.0
1 ppm (1 µg/g)	70-120	10	20	0.5-2.0
100 ppm (0.1%)	75-115	5	10	0.5-2.0
1%	85-110	3	5	0.5-2.0
10%	90-95	3	3	0.5-2.0
100%	95-105	2	3	0.5-2.0
		Microbiology		
Fals Positives (%)	False negatives (%)	Sensitivity (%)	Specificity (%)	Limit of detection (cfu/g or ml)
<9.6	<2	>97	>90	<3

*Note: for single laboratory validation the expected HORRAT would be 0.25-1.0.
ppb = parts per billion, ppm = parts per million, TBD = to be determined, cfu = colony-forming units.

Table 9.5. Suggested acceptance criteria for single laboratory validation (10, p. 185)

Concentration	Repeatability CV (%)	Reproducibility CV (%)	Mean recovery (%)
<1 µg/kg	36	54	50-120
1 µg/kg–0.01 mg/kg	32	46	60-120
0.01 mg/kg–0.1 mg/kg	22	34	70-120
0.1 mg/kg–1 mg/kg	18	25	70-110
>1 mg/kg	14	19	70-110

CV = coefficient of variation.

GENERAL DISCUSSION

The AOAC, over a period of more than 100 years, has generated the data needed, through collaborative studies, to be able to predict one of the key performance characteristics of a method, the relative standard deviation between laboratories (RSD_R). By examining the data from a large number of such studies, Horwitz showed in 1980 that, if one plotted the coefficient of variation vs. concentration, the results from individual analyses fit into a predictable pattern, the so-called Horwitz horn (14). The surprising conclusion was that one could confidently predict the RSD_R to be expected from an analysis based solely on the concentration of the analyte and largely independent of matrix, type of analyte, and type of method. It was later shown that one could quantify this relationship with the Horwitz equation.

$$RSD_R = 2C^{-0.1505}$$

where C is in terms of a decimal fraction (15). Using this relationship, one can calculate that at 1μg/g (C = 10^{-6}) the RSD_R should be about 16%, whereas at 1 ng/g (C = 10^{-9}) one would expect the precision to be about 45%, etc. Hall and Selinger characterized this relationship as "one of the most intriguing relationships in modern analytical chemistry" (16). Using this relationship, it can easily be shown that the RSD_R will increase by a factor of 2 for each decrease in analyte concentration of 2 orders of magnitude.

Comparing the RSD_R calculated by the Horwitz equation with that obtained in a collaborative study has proven to be a convenient way to evaluate the acceptability of a method. This is done by calculating the Horwitz ratio (HORRAT).

$$HORRAT = RSD_R(found)/RSD_R(calculated)$$

Experience has shown that HORRAT values bracketing 1.0 indicate a method with acceptable precision and clearly under statistical control. A HORRAT value of >2 indicates the method is not acceptable— i.e., it leads to poor results in the hands of experienced analysts in well-run laboratories. High HORRAT values may also indicate that: (a) the method as written is open to misinterpretation, (b) the samples distributed to the collaborating laboratories may have been nonhomogeneous, or (c) the laboratories involved may have had problems with standard preparation or stability. Low HORRAT values (<0.5) may indicate a lack of independence—i.e., unreported consultation between participants, replication by analysts until results are in agreement, or excessive rounding. On the other hand, low HORRAT values may also indicate the use of advanced technology, careful training of the analysts in conducting the method, use of common reference standards, and rigid adherence to a quality assurance program by all the laboratories involved in the collaborative study. The HORRAT concept is generally not applicable to empirical methods (moisture, ash, fiber determinations, etc.) and indefinite analytes (fiber, polymers, enzymes, etc.) and at the extremes of concentration (near zero as in the case of dioxins and 100% such as moisture determinations) (17).

REFERENCES

(1)   ISO (1999) *General requirements for the Competence of Testing and Calibration Laboratories*, ISO 17025, International Standards Organization, Geneva, Switzerland

(2)   ISO (1994) *Quality-Vocabulary*, ISO 3402, International Standards Organization ,Geneva, Switzerland

(3)   *The Fitness for Purpose of Analytical Methods. A Laboratory Guide to Method Validation and Related Topics*, Eurachem, Caparica, Portugal http://www.2.vtt.fi/82/ket/eurachem.html

(4)   ISO (1994) *Accuracy (Trueness and Precision) of Measurement Results, Pt. 1–6* ISO 5725, International Standards Organization, Geneva, Switzerland

(5)   CITAC (2000) *Quantifying Uncertainty in Analytical Measurement* 2nd Ed., Guide 3, Eurachem, Caparica, Portugal http://www.vtt.fi/ket/eurachem/quam2000-pl.pdf

(6)   CITAC (2001) *Guide to Quality in Analytical Chemistry*, Eurachem, Caparica, Portugal, http://www.vtt.fi/ket/eurachem/quam2000-pl.pdf

(7)   AOAC (2001) *Accreditation Criteria for Laboratories Performing Food Microbiological and Chemical Analyses in Foods, Feeds, and Pharmaceutical Testing*, AOAC INTERNATIONAL, Gaithersburg, MD

(8)   Garfield, F.M., Klesta, E., & Hirsch J. Eds (2000) 3rd Ed., *Quality Assurance Principles for Analytical Laboratories*, AOAC INTERNATIONAL, Gaithersburg, MD

(9)   Horwitz, W. (1995) *Pure Appl. Chem.* **67**, 331 & *J. AOAC Int.* **78**, 143A

(10)   Fajgelj, A., & Ambrus, A. (2000) *Principles and Practices of Method Validation*, The Royal Society of Chemistry, Cambridge, UK

(11)   Youden, W.J., & Steiner, E.A., Eds (1975) *Statistical Manual of the AOAC*, AOAC INTERNATIONAL, Gaithersburg, MD

(12)   Thompson, M., Ellison, S.L.R., & Wood, R. (2002) *Pure Appl. Chem.* **74**, 835

(13)   See the AOAC web site for detailed descriptions: http://www.aoac.org

(14)   Horwitz, W., Kamps, L.R., & Boyer, K.W. (1980) *J. Assoc. Off. Anal. Chem* **63**, 1344

(15)   Horwitz, W., & Albert, R. (1997) *Analyst* **122**, 615

(16)   Hall, P., & Selinger, B. (1989) *Anal. Chem.* **61**, 1465

(17)   Horwitz, W., personal communication

10. ANALYSIS OF MYCOTOXINS

Joerg Stroka and Elke Anklam

Food and Feed Unit
European Commission Joint Research Center
Institute for Reference Materials and Measurements
Retieseweg, B-2440
Geel, Belgium

Because of the unpredictable occurrence of mycotoxins in food and feed, analytical screening methods are necessary to identify mycotoxins and to assess whether a particular batch of material complies with legislation or trade regulations or whether it is contaminated at a level that is of concern for human or animal health. Appropriate analytical procedures combined with adequate sampling are a prerequisite for official food control. Since the discovery of mycotoxins and their potential threat to animal and human health, analytical methods have continuously been developed. The first available official methods were based on thin-layer chromatography (TLC). Because of developments in the field of instrumental analysis, liquid chromatography as well as enzyme-linked immunosorbent assay (ELISA) methods have been commonly available since the 1980s.

This chapter focuses on the development of analytical procedures, assuming that adequate sampling has been done and that homogeneous sample material is available for analysis.

GENERAL CONSIDERATIONS FOR METHOD SELECTION

Several considerations must be made for the selection of an analytical method. One of the most important aspects is of a legal character. As a result of different maximum limits set by various legislations worldwide (e.g., European Union Law, Codex Alimentarius Standards), the applicability of a method can be limited by a low legal limit depending on the method's limit of quantification (LOQ). One avoids operating at the LOQ of a method at which the uncertainty is large. Therefore, the method selected should have a LOQ considerably lower than the regulatory control level.

Another important consideration is linked to the performance requirements of a method. There is a significant difference between an analysis performed to screen goods for internal quality control and when the result must be defended in a court case (official control). In the first case, a certain number of false-positive results or limited information (e.g., whether a threshold is exceeded or not) is acceptable as long as the method offers other advantages such as rapid results. In the second instance (official control), results have a legal character and thus must be reliable and defendable. For this reason, the uncertainty (or certainty) of the result must be well defined (e.g., use of a reference method such as the AOAC or the European Standardization Committee method). However, these quantitative methods are often more time consuming. Further questions arise from laboratory and analysis management considerations, because some methods are well suited for single analysis of a sample (e.g., TLC), whereas others such as ELISA-based methods are economical only when large numbers of samples (n > 20) are analyzed simultaneously.

In addition to personal protection measures from exposure to the toxic mycotoxins, the analyst has to avoid degradation of the mycotoxin in the samples or standards during analysis. Degradation of the mycotoxin in the sample would lead to low (possibly false) results, and degradation of the analytical standards would lead to higher (false) results. Even though most mycotoxins are known to be fairly stable to most common food processing steps (e.g., cooking, frying), they can easily undergo degradation during analysis or storage for various reasons. Thus, improper procedures (e.g., exposure to daylight, storage of extracts in the wrong solvents, alkaline or acid atmosphere) can lead to significant degradation. For example, aflatoxins can be destroyed when they are exposed to daylight, to ultraviolet (UV) light (the exposure of an aflatoxin solution in transparent glass for a few minutes in a sunny place is sufficient for major degradation), or to an alkaline environment. An experienced analyst often becomes aware of degradation when new peaks appear in a chromatogram or when the pattern of peaks (e.g., in a mixture of aflatoxins B_1, B_2, G_1, and G_2) in standard solutions changes. The effect of daylight and the use of non-acid-washed glassware on the aflatoxins has been shown in a collaborative trial (1).

Fumonisins or ochratoxin A can degrade even without exposure to light or alkaline conditions. The content of spiked fumonisin in a sample matrix was shown to decrease without exposure to external factors (2). It was found that the matrix itself (fumonisins may react with residues of glucose in the matrix) could be responsible for degradation. Thus, it must be realized that fortified samples (e.g., for recovery calculation) are not stable for long periods of time. Another observation was that fumonisins could undergo degradation in certain solvents. For instance, fumonisins were more stable in acetonitrile/water than in methanol/water solutions (2). The same precautions must be taken with regard to ochratoxin A. This mycotoxin is also known to be susceptible to degradation and it is often required that glassware be treated with silane before use.

SPECIFIC TECHNICAL CONSIDERATIONS

History of Mycotoxin Analysis

Analytical methods for mycotoxins date back to the period when aflatoxins were discovered in the 1960s. At that time, TLC was applied as "the" method for identification and quantification; liquid chromatography and ELISA methods were not available. This matching discovery of a certain mycotoxin and the technology explains, to some extent, the number of TLC methods that have been published for certain mycotoxins (e.g., aflatoxins). However, for mycotoxins that were identified more recently to be a health threat to humans and animals (e.g., trichothecenes and fumonisins), methods based on liquid chromatography, gas chromatography, or ELISA were mostly developed first, because these compounds do not exhibit strong fluorescence or strong UV absorption.

Another important methodological aspect of chromatographic analysis methods is chemical derivatization. Generally, derivatization is an undesirable procedure for most analysts because it is often necessary to use aggressive and toxic substances. Nonetheless, in certain cases, such as when mycotoxins have no chromophores (no specific chemical structure that allows a sufficient absorption or fluorescence detection) or are difficult to separate by a certain technique (e.g., nonvolatile substances by gas chromatography), a derivatization step cannot be avoided (e.g., for fumonisin detection). As a result, the analyst might select a method that involves derivatization to suitably quantify a specific toxin.

Other aspects of method selection concern environmental issues, which are becoming more and more important. For example, chlorinated solvents such as chloroform and dichloromethane are considered to be hazardous to the environment and to human beings. Therefore, the aim is to substitute these solvents in state-of-the-art analytical procedures.

Sample Preparation and Cleanup

A prerequisite of any analytical method is proper sample cleanup. This stage of the analysis procedure is crucial, because sample cleanup can contribute to large systematic and random errors. Important factors are the particle size of the material, the appropriate extraction solvent, extraction mode, and extraction time.

Generally, the particle size of the material should be as small as possible. The practice of using only a small fraction of the material for analysis, which is common in other fields of analytical chemistry, is not recommended. Depending on the type and nature of the material to be analyzed, various sample preparation protocols have been stipulated.

Where practical, the entire quantity that is intended for human consumption must be milled sufficiently and used for analysis without any selective separation (e.g., sieving).

Several different extraction solvent systems have been proposed in the past depending on the mycotoxin being analyzed. Generally, these extraction solvents are based on mixtures of water and organic solvents such as methanol, acetone, and acetonitrile (mainly recently developed methods); chloroform; or dichloromethane. As already mentioned, there is an inclination to ban chlorinated organic solvents from daily laboratory use (which does not pose too many problems because in most cases they can be replaced by alternatives). Nonchlorinated extraction solvents are especially favored in methods that use immunoaffinity columns.

Furthermore, it must be remembered that the choice of the right extraction solvent can also serve as a selective sample cleanup procedure that allows the mycotoxin to be extracted and separated from matrix components that are less soluble in a specific extraction solvent and might interfere during the analysis. For example, aflatoxins are often extracted from animal feed with a mixture of acetone and water, because citrus pulp—an ingredient often used in animal feed—can lead to interference in chromatographic separations when the matrix is extracted with methanol and water. Such interference might not occur when the extract is analyzed by nonchromatographic techniques such as ELISA, and methanolic extraction solvents generally offer a better antibody stability.

Regarding the extraction efficiency of different solvent mixtures, it has been shown that the water/organic solvent ratio as well as the sample material/extraction solvent volume ratio plays an important role in method performance (e.g., recovery) (3–7). Many studies have been carried out to determine the best extraction conditions, and, as a result, recent methods often have similar extraction conditions. It must be stressed that any change in the extraction conditions is likely to lead to a change in method performance. Some interesting observations about aflatoxin analysis have been published, showing that the interaction of some extraction solvents (acetonitrile and water) with the sample matrix (e.g., spices such as paprika powder, infant formula, compound feeding stuff) can lead to false high recoveries (7). This must be noted, because not all extraction solvents are suitable for all matrices.

Slightly acidic conditions may be required during extraction of mycotoxins that contain a carboxylic acid group (R-COOH) to reduce deprotonation and thus improve extraction by the aqueous organic solvent mixture. In some cases, the addition of water-immiscible solvents such as hexane and cyclohexane is proposed when fatty matrices such as peanut butter are being analyzed. Such solvents have a purely physical effect on the extraction, allowing extraction of the mycotoxin from particles that are 'enclosed' in fat globules. In addition to the possibility of using various extraction solvents, there are various extraction modes, such as shaking and blending. These procedures increase the transfer speed of the mycotoxin from the matrix into the extraction solvent and have their specific applications. Shaking allows the simultaneous extraction of several samples (e.g., when using a horizontal shaker) without any cross-contamination, whereas blending (e.g., Ultra-Turrax) requires proper cleaning of the instrument between samples, but it is essential for matrices that would not mix sufficiently with the extraction solvent by shaking, even for a longer time (e.g., peanut butter).

After the extraction of the mycotoxin, a further purification step is required for most analytical methods. Generally, a superior cleanup is the prerequisite for sensitive methods with sufficiently low limits of detection (LOD), as required nowadays. In some early methods, which were designed for screening culture media in which high levels of mycotoxins could be found, or in those that were designed as multianalyte methods (simultaneous analysis of various mycotoxins in one run), a single mycotoxin-selective sample preparation step was skipped. The advantage of simultaneously determining several mycotoxins of different chemical structures is, as general rule, diminished by reduced sensitivity for each mycotoxin.

Official food control methods, with their demand of low LODs and LOQs, mostly require further sample cleanup procedures when based on chromatography. Exceptions are ELISA and related techniques already specific to mycotoxins in the crude extract.

The cleanup procedures described in recent methods are based on solid-phase extraction (patulin, fumonisins), immunoaffinity cleanup columns (IAC) (aflatoxins, fumonisins, ochratoxin A, deoxynivalenol, zearalenone), or on liquid-liquid separation (patulin in apple juice). The solid phase can either trap the mycotoxin, allowing the matrix interference components to be washed off, or trap the interfering substances.

Among these cleanup procedures, IAC has been shown to be the most powerful. When IAC is used appropriately, purified extracts can be obtained that result in clean chromatograms similar to those achieved from pure standard solutions. Even though IAC is still considered to be an expensive procedure, it is the cleanup procedure of choice, especially for difficult matrices that contain pigments (paprika, pistachios) or other interfering substances. In addition, each analyst should consider all the advantages and disadvantages, such as savings in time, chemicals, and overall efficiency as well as costs, before deciding on the most suitable method for sample cleanup procedure.

The most critical parameter for the IAC techniques is the amount of organic solvents in the extract that is to be purified. Generally, immunoaffinity columns contain antibodies that are immobilized on a support. These antibodies are subject to denaturation when exposed to higher concentrations of organic solvents (8), acidic conditions, or alkaline conditions, or to other unsuitable ambient parameters such as heat. As a matter of fact, immunoaffinity columns for aflatoxin analysis are to date more stable in methanol solutions (up to 15–20%)

than in acetonitrile and acetone (up to 2.5%) solutions. The sensitivity depends strongly on the type of antibody used. In most cases, IAC manufacturers make recommendations for their columns, which must be followed to ensure reliable results.

A further recommendation for the IAC procedure is to filter the extract through glass fiber filters under slight vacuum after dilution with water or phosphate-buffered saline and before the IAC stage. Filtration not only removes particles that cause clogging in the column, it also removes gases that are dissolved in the diluted extract. These gases often lead to microbubbles in the immunoaffinity gel and can also block the columns. Experience shows that filtration (even of clear extracts) in most cases allows the application of the extract by gravity, whereas without prior filtration, vacuum manifolds must be used to maintain a constant flow. The advantage of the gravity application is that the column does not run completely dry because the force is very small.

The reuse of immunoaffinity columns has been suggested. Manufacturers do not normally give any guarantee for results obtained with reconditioned immunoaffinity columns. However, a recent inter-laboratory comparison study, including the first author's laboratory, showed that reconditioning can be applied successfully under certain conditions. One of the most important parameters appears to be that the reconditioning has to be performed gently, while all solutions used (storage and reconditioning) are kept sterile and must be preserved with a suitable preservative that does not interact with proteins. Sodium azide was found to be such a preservative.

In solid-phase extraction, the most critical parameter is saturation of the solid-phase material. Because it is not as specific as the IAC material, it can be overloaded by applying a large amount of extract. Therefore, solid-phase extraction methodology requires the analyst to carry out the steps of a validated official method strictly, without modification.

Enzyme-Linked Immunosorbent Assay

This technique requires no special instrumentation apart from appropriate plate readers. Enzyme-linked immunosorbent assay kits are available for a wide range of mycotoxins (technically, for any substance for which an antibody is available). The advantage of ELISA methods is that they allow simultaneous analysis of a large number of samples within a reasonably short time. However, any cross-reaction between the ELISA antibodies and interfering substances can lead to false-positive results and may require reanalysis by a different method (e.g., TLC or liquid chromatography). Another limitation is that ELISA methods can determine the total amounts of only one mycotoxin class (e.g., total aflatoxins or total fumonisins). It should be noted that, in some cases, legislation stipulates regulation of a single mycotoxin amount within the same class (e.g., aflatoxin B_1 or fumonisin B_1 only). Nevertheless, the ELISA technique is a very valuable tool for screening large numbers of samples at minimal cost. An overview of immunoassay methods for mycotoxins is provided in chapter 11. Positive samples may require confirmation by other techniques such as quantitative official TLC or liquid chromatography.

Liquid Chromatography and Gas Chromatography

Liquid and gas-chromatographs are by far the most unique and widely applied instruments, and the methods in general are considered reliable for many types of trace analyses. Their high resolution, in addition to efficient cleanup procedures and selective detection methods, offers a high degree of confirmative (identification) and quantitative information. Because of the variety of different detectors available for both methods (e.g., fluorescence, UV absorption, electrochemical or mass-spectroscopy for liquid chromatography and flame ionization; electron capture or mass spectrometry for gas chromatography) as well as the general and steady progress in instrumental development, these methods are currently used for a variety of mycotoxins and other substances. Currently, liquid chromatography is the most commonly used method for mycotoxin analysis. However, both gas chromatography and liquid chromatography require a high level of maintenance and have special demands on the purity of chemicals used (gases or solvents) and the stability of the power supply. Today's instruments are normally computer controlled and require more laboratory space than TLC or ELISA. Instrumental chromatographic techniques are the methods of choice in regions with sufficient infrastructure, because a single instrument is normally used for different types of analysis in a service laboratory instead of for a certain type of analysis only.

Thin Layer Chromatography

Thin layer chromatography is by far the most simple and thus appropriate method when no special or expensive laboratory instrumentation is available. Thin layer chromatographic separation is preferably performed on silica gel plates. Generally, chromatograms can be evaluated visually for several important mycotoxins because of their fluorescence and colored derivatization products (aflatoxins, ochratoxin A, fumonisins). It must be realized that only mycotoxins with an intrinsic fluorescence, or those for which adequate derivatization methods are available, can be directly determined visually at sufficiently low levels. The disadvantage of visual inspection is its strong dependence on the visual ability of the analyst. Generally, most analysts (even when they are well trained) encounter difficulties distinguishing intensities of less than a 20% difference.

Densitometry is the second most commonly used method for mycotoxin determination after TLC separation: the first is visual determination. As an economical alternative to commercially available TLC densitometers, two new approaches applying cheap and simple instruments that have been developed recently have been shown to be suitable for determining aflatoxins (8, 9). These new techniques are currently under evaluation in an interlaboratory comparison study. Two-dimensional TLC or bidirectional development of the plate is often used when the mycotoxin cannot be sufficiently purified during cleanup and the development by a single mobile phase is not sufficient to separate the toxin during the chromatographic run. The disadvantage of this technique, however, is that the number of samples per plate is low (generally one sample and one standard per plate). On the other hand, sufficiently selective sample cleanup procedures (e.g., immunoaffinity) before TLC can allow several samples and standards to be simultaneously determined on a single plate (one-dimensional development) (10). High-performance TLC (plates with a particle size of about 3 μm) offers the advantage of small and dense spots, resulting in higher intensity and lower detection limits.

An interesting aspect that should be investigated is the possibility of reusing certain TLC plates. So far there is neither anecdotal evidence nor published literature that shows reuse of ordinary TLC plates in mycotoxin analysis. However, reusable TLC plates (so called *Permakote* plates) are available and have been used. It has been reported that their performance is not diminished when the plates are reconditioned several times. Permakote TLC plates are made of sintered silica gel and contain no binder that would be destroyed during the reconditioning process with sulfuric acid /dichromate solution.

Fluorimetry

Fluorimetric methods use fluorescence measurement on minicolumns or in solution and are a simple alternative for rapid mycotoxin determination. Such methods have been developed in the past, especially for aflatoxins (11, 12). Recently, some interest in the use of fluorimetry in solution to determine various mycotoxins (e.g., fumonisins, zearalenone) has been shown. Solution fluorimetry methods do not differ from the previously discussed chromatographic methods with regard to sample extraction and cleanup procedures. When IAC is required for solution fluorimetry of mycotoxins, disadvantages similar to those already discussed for the ELISA methods are encountered. In addition, fluorimetry on silica gel columns has so far been done only by visual inspection; therefore, the method is subject to uncertainties.

Less Common but Novel Techniques

Because of the growing interest in mycotoxin analysis, several additional methodologies have been developed. These technologies range from very simple-to-use methods (*dipstick* tests) (13) to high-technology instrumentation methods. The common principle of all these methods is that they involve immunochemistry. In fact, the availability of mycotoxin-specific antibodies was the principal catalyst for the advances in the whole field of mycotoxin analysis. Dipstick technology allows the analyst to check for the presence of a mycotoxin in a simple way, similar to blood glucose and pregnancy tests. An interesting approach that is already available commercially is that based on a simple membrane, which can be checked visually and does not require volumetric glassware. This method has been recently described (13), although the authors did not indicate performance parameters. Other currently available high-tech techniques are so-called fluorescence polarization (14), surface plasmon resonance, and another immunochemistry-related technology based on the ORIGEN technology, which uses electrically induced chemoluminescence for detection. These technologies cannot be described here but are mentioned because they appear to be promising approaches for the future. Until now, only fluorescence polarization has been applied for determining mycotoxins (15). However, it must be mentioned that all these methods are still under development.

AVAILABILITY OF OFFICIAL ANALYTICAL METHODS

One of the most critical barriers for many analysts is tracking sources for suitable analytical methods (the large number of scientific publications complicates the process of selecting a suitable method). In addition, the rapid development of new methods makes it difficult to maintain one state-of-the-art standard method, if necessary. Recourse may be found, however, in the several standardization organizations that supply suitable validated methods for official analyses of food and other materials for various analytes. The *Official Methods* book of AOAC INTERNATIONAL and the European Standardisation Committee's validated methods are available to analysts.

When no "official methods" are available or needed, the analyst is advised to refer to review articles in peer-reviewed journals or to scientific books as a source for suitable methods, procedures, and techniques. A selection of current relevant sources is listed in the references section (15–17). Additional information and material that became available lately for analysts are training videos and method descriptions for the most relevant mycotoxins that are available on the Internet (18) or as a CD-ROM (19). These media are intended to familiarize analysts with the most common procedures and critical steps in mycotoxin analysis.

SUMMARY

This paper has described and compared currently available analytical methods for determining the most relevant mycotoxins. The methods and their principles have been discussed with regard to their applicability for food monitoring in terms of legal requirements, practicability, simplicity, and method performance. Besides general considerations for method selection and specific technical considerations, a special focus was put on appropriate methods for monitoring mycotoxins as early as possible in the production of agriculture products and in the food processing chain.

REFERENCES

(1)   van Egmond, H.P., Heisterkamp, S.H., & Paulsch, W.E.E. (1991) *Food Addit. Contam.* **8**, 17

(2)   Viscont, A., Doko, M.B., Bottalico, C., Schurer, B., & Boenke, A. (1994) *Food Addit. Contam.* **11**, 427

(3)   Bradburn, N., Coker, R.D., & Blunden, G. (1995) *Food Chem.* **52**, 179

(4)   Bradburn, N., Coker, R.D., Jewers, K., & Tomlins, I.K. (1990) *Chromatographia* **29**, 435

(5)   Chang, H.H.L, de Fries, J.W., & Hobbs, W.E. (1979) *J. Assoc. Off. Anal. Chem.* **62**, 1281

(6)   Shotwell, O.L., & Goulden, M.L. (1977) *J. Assoc. Off. Anal. Chem.* **60**, 83

(7)   Stroka, J., Joerissen, U., Petz, M., & Anklam, E. (1999) *Food Addit. Contam.* **16**, 331

(8)   Scott, P.M., & Trucksess, M.W. (1997) *J. AOAC Int.* **80**, 941

(9)   Stroka, J., & Anklam, E. (2000) *J. Chromatogr. A*, **904**, 263

(10)   Stroka, J., van Otterdijk, R., & Anklam, E. (2000) *J. Chromatogr. A* **904**, 251

(11)   AOAC (2000) *Official Methods of Analysis*, 17th Ed., AOAC INTERNATIONAL, Gaithersburg, MD, Method 975.36

(12)   AOAC (2000) *Official Methods of Analysis*, 17th Ed., AOAC INTERNATIONAL, Gaithersburg, MD, Method 979.18

(13)   Schneider, E., Ursleber, E., & Maertlbauer, E. (1995) *J. Agric. Food Chem.* **43**, 2548

(14)   De Saeger, S., & Van Peteghem, C. (1999) *J. Food Prot.* **62**, 65

(15)   Maragos, C.M., Jolley, M.E., Plattner, R.D., & Nasir, M.S. (2001) *J. Agric. Food Chem.* **49**, 596

(16)   Betina, V. (1989) *J. Chromatogr.* **477**, 187

(17)   Betina, V. (1985) *J. Chromatogr.* **334**, 212

(18)   http://cpf.jrc.it/mycotoxin/index.htm

(19)   elke.anklam@jrc.it

11. IMMUNOASSAY PROCEDURES FOR SCREENING FOOD COMMODITIES FOR MYCOTOXINS: AN OVERVIEW

Henry Njapau

Center for Food Safety and Applied Nutrition
Food and Drug Administration
5100 Paint Branch Parkway
College Park, Maryland 20740

The natural occurrence of mycotoxins in food crops is sporadic and the distribution of for instance aflatoxin B_1 within a lot of contaminated material is usually uneven. Furthermore, the incidence of contamination is highly dependent on multiple factors and will vary over a period of time even within a specified geographical area. Contamination may be confined to a small portion of the lot increasing the likelihood of misrepresenting the mean value of a mycotoxin in the lot. By increasing the number of samples assayed, the mean result of the multiple analyses gets closer to the true value. Therefore, in order to obtain valuable survey or monitoring data, it is necessary to test large numbers of samples hence requiring assay methodology that can be performed rapidly and with minimum technical dexterity. The rapid processing of a large number of samples in order to identify those requiring confirmatory application with a rigorous method is called screening (1). Screening for mycotoxins has been performed using the Blue Greenish-Yellow Fluorescence (BGYF) test, minicolumns, bioassays, visual inspection and immunoassays. Immunoassays possess the most desirable characteristics of a screening method, offer the advantage of high specificity and a reduction in assay time, and most can be used for quantitative analysis as well. This section presents a brief and simple overview of immunoassays and their use for mycotoxin analysis. It is not a comprehensive treatise on immunochemistry. For in-depth review of immunoassay methodology the reader is referred to other sources (2-7).

Since the reported use of antibodies for quantitative analytical purposes in 1932 (8), immunoassays have been used in a wide range of fields that include, diagnostic medicine (9), and pesticide (10), aquatic biotoxin (11), fungal (12) and mycotoxin analysis. Pioneering work regarding the application of immunoassay technology to mycotoxin analysis was performed by Chu and coworkers (13). Subsequently, several immunochemical assays based on polyclonal (14) and monoclonal (15) antibodies were developed for various mycotoxins. The performance of the initial immunochemical methods developed for mycotoxin analysis (14, 16, 17) was described as comparable to conventional techniques such as thin layer chromatography.

PRINCIPLES OF IMMUNOASSAYS

Immunoassay technology is founded on the basic antigen-antibody reaction. When higher vertebrates are exposed to material of foreign origin, antibodies are elicited as part of the defense mechanism of the animal. Antibodies are proteins produced by bone marrow immunocytes called B-lymphocytes (white blood cells) in response to the antigen challenge. Several cells are often stimulated in the same animal, each synthesizing an antibody specific to a distinct portion of the antigen. Since numerous original cells are involved, the product is termed polyclonal. A single antibody-producing cell can be isolated and grown or multiplied *in vitro* to produce several genetically identical cells that will produce one type of antibody classified as monoclonal (18). Monoclonal antibodies exhibit a much higher degree of specificity. In order to elicit an immune response, however, the antigen must be sufficiently large, of the order of 10,000 molecular weight units (17). Aflatoxins (with a molecular weight of about 320) are too small to elicit an immune response on their own. Small molecules not antigenic by themselves (haptens) can stimulate the production of antibodies, if conjugated to a larger molecule, usually a protein carrier (14).

The binding site on an antibody is complimentary to a minor structure or portion of the antigen-carrier conjugate (epitope) and not the whole complex (19). The epitope is usually a three dimensional conformation of a portion of the hapten (e.g. aflatoxin) that triggered the formation of the antibody. In the aflatoxin molecule, the epitope may be the dihydrofuran moiety or the lactone ring. Hence, the site of conjugation of the aflatoxin molecule to the protein carrier is important because it influences the location of the epitope. For instance, the majority of aflatoxin B_1 transformations occur through the dihydrofuran moiety. If the sample matrix possesses active metabolic enzymes, the dihydrofuran region of the aflatoxin molecule may be modified and antibodies specific for this region may not recognize the modified structure thereby rendering the assay ineffective.

Furthermore, when the epitopic structural conformation is present on precursors, analogs or metabolites, cross-reactivity may occur. In the case of the structurally analogous aflatoxins some level of cross reactivity has been shown, particularly when the dihydrofuran ring is or contributes to the epitopic structure. Consequently, antibodies raised against aflatoxin B_1 react with the other aflatoxins. Table 11.1 shows percent cross reactivity of antibodies produced against aflatoxin B_1 with the other aflatoxin analogs.

Table 11.1. Percent cross reactivity of anti-aflatoxin B_1 antibodies with the various aflatoxin analogs

Antibody type	Percent cross reaction with				Reference
	AFB_1	AFB_2	AFG_1	AFG_2	
Polyclonal	100	4	1	0.01	(31)
Polyclonal	100	28	100	22	(32)
Monoclonal	100	100	5	4	(28)
Monoclonal	100	20	34	17	(33)
Monoclonal	100	20	25	–	(18)

IMMUNOASSAYS FOR MYCOTOXINS

Until recently, there were three basic types of immunochemical methods: radioimmunoassay (RIA), enzyme-linked immunosorbent assay (ELISA) and immunoaffinity columns (IAC). Potential health hazards, waste disposal problems, and half-life concerns have virtually eliminated the use of radioimmunoassay (20) in mycotoxin analysis. In its place is an array of novel immunotechniques ranging from disc-based microarrays (21) to chemiluminescence (22) and flow-through systems (23). ELISA methods are available for the detection of several mycotoxins (aflatoxins, the trichothecenes, zearalenone, ochratoxins, fumonisins and deoxynivalenol) in cottonseed, barley, wheat, peanuts, corn and aflatoxin M_1 in milk (17). They typically employ an extraction into aqueous methanol followed by direct dilution in a buffer, offering significant advantages in terms of reduced assay time. Depending on the purpose and type of equipment available, these assays may be used for qualitative, semi-quantitative or quantitative analysis.

ELISA Procedures

Enzyme-linked immunosorbent assays may be set up based on a competitive or non-competitive format. Competitive assays attain maximal sensitivity with minimal amounts of antibody whereas excess antibody over the antigen is used in the non-competitive formats. Within each category, the assay protocol may be direct or indirect (Table 11.2). For most mycotoxins, the commonly used versions are the direct and indirect competitive ELISAs. In direct competitive ELISA, the microtiter plate wells are coated with an antibody that recognizes the mycotoxin. The sample extract and an enzyme-bound toxin are simultaneously added to the wells. The enzyme-bound toxin competes with free toxin from the sample for the antibody in the wells. The antibody reacts more readily with the free toxin than the enzyme-bound toxin. Excess free toxin and/or enzyme-bound toxin are washed out and a chromogen (substrate) is added. The substrate reacts with the enzyme on the enzyme-toxin conjugate forming colored product.

The color reaction is stopped by the addition of buffered dilute acid (24). Since the amount of substrate that is hydrolyzed is directly proportional to the quantity of enzyme available, the intensity of the color is inversely proportional to the amount of toxin in the sample extract.

Table 11.2. Summary of principal steps and characteristics of the common versions of ELISAs

Competitive		Non-Competitive	
Direct	Indirect	Direct	Indirect
• Microwell coated with antibody	• Microwells coated with toxin	• Microwell coated with antibody	• Microwell coated with toxin
• Sample extract + toxin-enzyme conjugate	• Sample extract + first antibody	• Sample extract only	• First antibody only
• Chromogen	• Second antibody-enzyme conjugate	• Antibody-enzyme conjugate	• Second antibody-enzyme conjugate
• Reaction stoppage	• Chromogen	• Chromogen	• Chromogen
	• Reaction stoppage	• Reaction stoppage	• Reaction stoppage

In indirect competitive ELISA, the microtiter plate wells are coated with a protein-mycotoxin conjugate (25). A sample extract and a mycotoxin-specific antibody are added to the wells and allowed to react. The protein-mycotoxin conjugate bound to the wells competes with the free toxin in the sample extract for the antibody. The antibody will preferentially bind the aflatoxin in the sample extract. When the sample extract contains no aflatoxin, the antibody will bind to the toxin bound to the microtiter wells. After incubation, the wells are washed and a second antibody (bound to an enzyme) is added. The second antibody recognizes and binds the first antibody, which, in the absence of free toxin in the sample, would be bound to the toxin in the microtiter wells. When a chromogen (or substrate) is added to the wells, the enzyme on the second antibody hydrolyzes the substrate to produce a colored product. The color reaction is stopped as in direct ELISA and the color of the solution in the microtiter wells is similarly inversely proportional to the concentration of toxin in the sample.

The presence of toxin in the sample may be determined visually based on the presence or absence of a color or the intensity of the color in the microwells in reference to a standard. The quantity of toxin may also be calculated from standard curve. Direct ELISA is one step shorter than indirect ELISA. However, the advantages of direct ELISA may be negated by observations that antibodies adsorbed directly to the solid phase (as in direct ELISA) reacted with the antigen (mycotoxin) less efficiently than those that are in solution as in indirect ELISA (26).

Immunoaffinity Columns (IAC) Assay Procedure

In the immunoaffinity column technique, an antibody is adsorbed on silica gel or Sepharose gel. The gel is suspended in a buffer solution and slurried into a column. A dilute sample extract is passed through the column by gravity or mechanical force. The toxin is captured by the antibody and retained on the column while most of the extract passes through. Washing the column with water or buffer elutes any weakly bound undesired matrix molecules. The anlyte (toxin) is eluted from the column with an appropriate solvent, usually absolute methanol or dimethyl sulfoxide (DMSO) (27, 28). The quantity of the toxin in the sample is estimated using a fluorometer. Sometimes the immunoaffinity column is used as an intermediate cleanup step for other analytical techniques such as high performance liquid chromatography (HPLC).

MYCOTOXINS IMMUNOASSAY TEST KIT SOURCES

Several companies offer rapid mycotoxin screening test kits in various forms. Some of the manufacturers are listed in Table 11.3. Only one reference address for each company has been included in the table. However, some of the companies have subsidiaries in more than one country or location. For instance, Romer Labs® has branches in the USA and Singapore in addition to the one in Austria.

Table 11.3. Some sources of rapid test kits for mycotoxin analysis

Source	Mycotoxins	Test Kit Brand Name	Format
Charm Sciences Inc., 659 Andover Street, Lawrence, MA 01843, USA Tel: +1 978 687 9200 Fax: +1 978 687 9216 www.charm.com	Aflatoxins Aflatoxin M1 Deoxynivalenol (DON) Ochratoxin Zearalenone	ROSA	Lateral flow devices (LFDs)
Diagnostix, 2845 Argentia Road, Unit 5, Mississuaga, Ontario, Canada LSN 8G6 Tel: +1 905 286 4290 Fax: +1 905 286 5260 www.diagnostix.ca	Aflatoxins DON Fumonisins Zearalenone	EZ-Quant	Microtiter (microwell) ELISAs
Indexx Laboratories Inc. One Idexx Drive, Westbrook, ME, USA Tel: +1 207 556 0300 Fax: +1 207 556 4346 www.idexx.com	Aflatoxin M1	SNAP	LFDs/ Immunodot ELISA
International Diagnostics Systems Corp., 2620 South 100, P.O. Box 799, St. Joseph MI, USA Tel: +1 269 428 8400 Fax: +1 569 428 0093 www.ids-kits.com	Aflatoxins	Afla-Cup Afla-Strip	Immunodot ELISA LFD
Neogen Corp 620 Lesher Place, Lansing, MI 48912, USA Tel: +1 517 827 4279 Fax: +1 517 372 2006 www.neogen.com	Aflatoxins DON Fumonisins Ochratoxin T-2 toxin Zearalenone	Agri-Screen Veratox Reveal NeoColumn	Microtiter ELISAs LFD Immunoaffinity Columna (IAC)
Raisio Diagnostics OY, Joukanaisenkatu 1, FL- 20520 Turku, Finland Tel: +358 2 241 0227 Fax: +358 2 241 0278 www.diffchamb.com	Aflatoxins DON Fumonisins Ochratoxin A Zearalenone	Transia	Microtiter ELISAs

R-Biopharm Rhone Ltd, West of Scotland Science Park, Unit 3.06 Kelvin Campus, Glasgow G20 0SP Scotland Tel: +44 (0) 141 945 2924 Fax: +44 (0) 141 945 2925 www.r-biopharmrhone.com	Aflatoxins DON Fumonisins Ochratoxin T-2 Toxin Zearalenone	AFLACARD AFLACARD- TOTAL AFLAPLATE AFLASCAN OHRACARD OCHRASCAN	Immunodot ELISA Microtiter ELISA IAC/UV column Immunodot ELISA IAC/UV column
R-Biopharm, Landwehrstr. 54 64293 Damstadt, Germany Tel: +49 6151 8100/1 Fax: +49 6151 8102/3 www.r-biopharm.com	Aflatoxins Aflatoxin M$_1$ DON Fumonisins Ochratoxin T-2 Toxin Zearalenone	RIDASCREEEN RIDASCREEN -FAST RIDAQUICK	Microtiter ELISA LFD
Romer Labs Diagnostic GmBH Technopak 1,3430 Tulln, Austria Tel: +43 2272 615331 Fax: +43 2272 615331-11 www.romerlabs.com	Aflatoxins Aflatoxin M1 DON Fumonisins Ochratoxin T-2 Toxin Zearalenone	Agra Strip Afla-Cup AgraQuant FluoroQuant AccuTox AflaStar MycoSep	LFD Immunodot ELISA Microtiter ELISA Fluorometric Immuno-tube Cleanup IAC Cleanup columns
Strategic Diagnostics Inc., 111 Pencader Drive, Mewark, Delaware, 19702, USA Tel: +1 302 456 6789 Fax: +1 302 456 6782 www.sdix.com	Aflatoxins DON Fumonisins	Myco-Chek	Microtiter ELISA
Vicam LP, 313 Pleasant Street Watertown, MA 02472, USA Tel: +1 617 926 7045 Fax: +1 617 923 8055 www.vicam.com	Aflatoxins Aflatoxin M$_1$ Citrinin DON Fumonisins Ochratoxin T-2 Toxin Zearalenone	AflaTest x*-Test	Cleanup IAC columns prior to fluorometric or HPLC analysis

(1) *x = Abbreviated toxin name (e.g. 'Afla' for aflatoxin and ending with the term 'Test')

SUMMARY

Immunoassays are rapidly becoming the method of choice where high sensitivity and throughput are desired. A wide range of qualitative and quantitative immunoassays tests have been developed based on the specificity of the antibody antigen reaction. The high specificity and affinity of the reactants allows the analysis of samples without exhaustive purification. In their simplest form, ELISA techniques are designed to rapidly give a visual indication of whether or not the toxin or analyte content of a sample exceeds a specified amount. This relative simplicity and inherent low cost makes them an ideal screening tool. The performance of immunochemical assays for mycotoxin analysis in corn, cottonseed, peanuts and peanut products and milk has been evaluated in collaborative and/or single laboratory studies (28, 29). The aflatoxins are determined as a sum parameter and total contamination in the 1-5 ppb (µg/kg) range is reliably detectable (22, 26, 27). Manufacturers of various immunoassay test kits indicate an even lower limit of detection, in parts per trillion (ppt, ng/kg) for various fluid matrices. Caution must however be exercised when handling immunoassay reagents or test kits. Antibodies and enzymes are proteins and may be inactivated under extremes of both humidity and temperature.

REFERENCES.

(1)   Horwitz, W., Cohen, S., Hankin, L., Krett, J., Perrin, C.H. & Thornburg, W. (1978) in *Quality Assurance Practices for Health Laboratories*, Inhorn, S.L. (Ed), American Public Health Association, Washington, D.C. 588.

(2)   Tjissen, P. (1985) *Practice and Theory of Enzyme Immunoassays*, Elsevier, Amsterdam, The Netherlands

(3)   Gosling, J.P. (1990) *Clin. Chem.* **36,** 1408-1427

(4)   Challocombe, S.J. & Kemeny, D.M. (1998) *ELISA and Other Solid Phase Immunoassays: Theoretical and Practical Aspects*, John Wiley & Son, Chichester, West Sussex, England, pp378

(5) Edwards, R. (1996) *Immunoassays: Essential Data*, John Wiley & Son, Chichester, West Sussex, England, 178 pp

(6) Warsinke, A., Benkert, A. & Scheller, F.W. (2000) *Fresenius. J. Anal. Chem.* **366**, 622

(7) Weller, M.G. (2000) *Fresenius. J. Anal. Chem.* **366,** 635

(8) Heidelberger, M. & Kendall, F.E. (1932) *J. Exp. Med.* **55,** 555

(9) Walsh G., O'Shaughnessy, B., Shanley, N. & Tobin, J.T. (1998) *Biochem. Educ.* **26,** 157

(10) Wang, S., Hill, A.S. & Kennedy, I.R. (2002) *Anal. Chim. Acta* **468,** 209

(11) Chu, F.S. & Fan, T.S. (1985) *J. Assoc. Off. Anal. Chem.* **68**, 13.

(12) Yong R.K. & Cousin M.A. (2001) *Int. J. Food Microbiol.* **65,** 27

(13) Chu, F.S. (1983) in *Proceedings of the International Symposium on Mycotoxins*, September 6-8, 1981, Cairo, Egypt, Naguib, K.H., Naguib, M.M., Park, D.L. &   Pohland, A.E. (Eds), National Research Center, Dokki, Cairo, Egypt, 177

(14) Chu, F. S. (1986) in *Modern Methods of Analysis and Structural Elucidation*. Cole, R.J. (Ed), Academic Press, New York, 294.

(15) Groopman, J.D., Cain, L.G. & Kensler, T.W. (1988), *CRC Cr. Rev. Toxicol.* **19,** 113.

(16) Shepherd, M.J., Mortimer, D.M. & Gilbert, J. (1987) *J. Assoc. Publ. Analysts* **25**: 129.

(17) Pestka, J.J. (1988) *J. Assoc. Off. Anal. Chem.* **71,** 1075.

(18) Kaveri, S. V.,Fremy, J-M., Lapeyre, C. & Strosberg, A.D. (1987) *Lett. Appl. Microbiol.* **4,** 71.

(19) Vaag, P. & Munck, L. (1987) *Cereal Chem.* **64,** 59

(20) Blake, C. & Gould, B.J. (1984) *Analyst* **109,** 533

(21) Kido H., Maquieira A., & Hammock, B.D. (2000) *Anal. Chim. Acta* **411,** 1

(22) Dodeigne, C., Thunus, L., & Lejuene, R. (2000) *Talanta* **51,** 415

(23) Ho, J. A. & Durst, R.A. (2000) *Anal. Chim. Acta* **414,** 61

(24) Park, D.L., Miller, B.M., Nesheim, S., Trucksess, M., Veckich, A., Bidigare, B., McVey, J. & Brown, L.H. (1989b) *J. Assoc. Off. Anal. Chem* . **72,** 638

(25) Chu, F.S. & Lee, R.C. (1988) *J. Assoc. Off. Anal. Chem.* **71,** 953

(26) Dierks, S.E., Buttler, J.E. & RIchardson, H.B. (1986) *Mol. Immunol.* **23,** 403

(27) Trucksess, M.W., Young, K., Donahue, K.F., Morris, D.K. & Lewis, E. (1990) *J. Assoc. Off. Anal. Chem.* **73,** 425.

(28) Groopman, J.D. & Donahue, K.F. (1988) *J. Assoc. Off. Anal. Chem.* **71,** 861

(29) Trucksess, M. W., Stack, M.E., Nesheim, S., Park, D.L. & Pohland, A.E. (1989) *J. Assoc. Off. Anal. Chem.* **72,** 957

(30) Lee, L.S., Wall, J.H., Cotty, P.J. & Bayman, P. (1990) *J. Assoc. Off. Anal. Chem.* **73,** 581

(31) Fan, T.S. & Chu, F.S (1984) *J. Food Prot.* **47,** 263

(32) Morgan, M.R.A., Kang, A.S. and Chan, H.W-S. (1986) *J. Sci. Food Agric.* **37,** 908

(33) Dixon-Holland, D.E., Pestka, J.J., Bidgare, B.A., Casale, W.L., Warner, R.L., Ram, B.P. & Hart, L.P. (1988) *J. Food Prot.* **51,** 201

12. THIN-LAYER CHROMATOGRAPHY IN MYCOTOXIN ANALYSIS

Mary W. Trucksess

Center for Food Safety and Applied Nutrition
Food and Drug Administration
5100 Paint Branch Parkway
College Park, Maryland 20740, USA

Thin-layer chromatography (TLC) became important for the separation of analytes in complex matrices not amenable to analysis by gas chromatography after the pioneering work of Justus Kirchner and Egon Stahl. The rapid growth of TLC slowed during the 1970's with the rise in popularity of high performance liquid chromatography (HPLC). Although both HPLC and TLC are capable of separating the same types of compounds, HPLC is considerably more reproducible for quantitation and retention values than TLC. However, the improvements in TLC have reduced or eliminated most of its limitations. In some cases it may give better resolutions and results than HPLC. It can be applied to separations and detections for which HPLC cannot be used. TLC is always cheaper and faster and can be applied with less sample extract cleanup.

Many of the improvements in TLC may be placed under the heading of high performance thin-layer chromatography (HPTLC). Originally, HPTLC was used to describe the use of small, uniform 5-20 μm particle adsorbents, in contrast to the 37-70 μm adsorbents first used. Instrumental quantitation in TLC has also been improved. The definition of HPTLC has now been broadened to cover a variety of approaches. Included are multidimensional, multisolvent, multidevelopment and totally instrumentalized procedures from sample extract application and densitometric scanning of the plate to collection and evaluation of the data by computer. TLC separation includes sample application, plate development, visual observation and quantitation.

TLC EQUIPMENT

The basic equipment needed for TLC can be low cost and include a sample applicator, TLC plate, developing tank containing the mobile phase, and a longwave ultra violet (UV) lamp for fluorescent analyses. The applicator can be a micro syringe or an automatic sampler. There are 2 types of plates: the standard TLC plate or simply referred to as TLC plate and high performance TLC plate (HPTLC). Standard TLC plate is usually 20 x 20 cm, 0.25 mm thick layer of 5-20 μm particles. The adsorbent can be normal phase or reverse phase. Normal phase TLC is more commonly used than reverse phase TLC for mycotoxins analysis. TLC normal phase adsorbents include silica gel, aluminum oxide and cellulose. Reverse phase (RP) absorbents are silica gel bonded with ethyl, octyl, octadecyl, phenyl, or cyano functional groups. The support may be glass, aluminum or plastic. During the past 20 years, a wide variety of high resolution HPTLC plates have become available. The stationary phase adsorbents have small particle sizes and narrow particle size distributions. The plates are usually 10 x 10 cm or 10 x 20 cm. The development tanks may be metal or glass. A longwave UV lamp is used for visual observation, estimation or quantitation of the mycotoxins and/or fluorescent derivatives spots. The TLC plate can also be quantitated by fluorodensitometry with a fluorodensitometer.

TLC PROCEDURE

In general the procedure consists of dissolving test extract residue, spotting this solution on the plate, preparing a mobile phase, developing the plate, observing the spots, and estimating or quantitating the chromatogram.

Application of Test Extract

The solubility of the remaining residue after purification is an important consideration in selecting the solvent for spotting. Purified test extract must be soluble in spotting solvent. Solvent must not spread analyte away from point of application. The spotting volume is kept to minimum but sufficient to contain desired quantity of analyte. Usually the volume is 1-50 μL. Narrow band application gives higher resolution and allows applying larger amounts but requires more time and reduces the number of tests per plate. Spot diameters should be approximately 3 mm and 1mm for traditional TLC and HPTLC, respectively, for best separation. A microsyringe or an automatic applicator provides accurate application volume. Compact spots are best obtained by a slow rate of test extract solution delivery. The application point (origin spot) of test extract solution should be close to the lower edge of the plate, usually about 2 cm, but above the developing solvent line to avoid wash out of the spots into the developing solvent. Damaging absorbent layer while spotting must be avoided. For quantitation, 4 or 5 spots of different amounts of standard analyte must be applied to the same plate as the test extracts.

Preparing the Mobile Phase

The retention factor (R_f) for a given type of absorbent varies depending on the choice of mobile phase. The choice of the mobile phase depends on the analyte, the absorbent, and test matrix. For normal phase development, a solution of a relatively larger amount of nonpolar solvents (chloroform, diethyl ether, and toluene) and a smaller amount of polar solvents (water, ethanol, acetone, and formic acid) is used. To improve resolution of mycotoxins from interfering compounds, usually a mixture of more than 2 solvents is used. In the RP-TLC system the mobile phase is more polar than the plate coating, therefore, larger amounts of the polar solvent (water containing 1-3% KCl) are mixed with smaller amounts of nonpolar solvents.

Developing the Plate

After spotting, the plate is placed in a glass or metal tank with the selected developing solvent. The developing tank must be sealed to avoid evaporation of volatile solvents and kept at a constant temperature.

The development distance is about 15 cm for TLC and about 7 cm for HPTLC. For difficult separation involving complicated matrices various development techniques can be used. Five types of linear ascending developments have been used for mycotoxins: one solvent, two solvents, bidirectional (BiD), automated multiple development (AMD) and 2 dimensional (2D). The one-solvent is self-explanatory. In the two-solvent development, the plate is first developed with a solvent that removes the interferences; then the plate is dried and developed with another solvent in the same direction to separate the toxins. In the bidirectional development, the test extracts are spotted in the middle of the plate. After the first development with a nonpolar solvent to remove the nonpolar components, the top of the plate below the solvent front is cut off. The plate is then turned upside down (180°) and developed with a more polar solvent to separate the toxins. In AMD or overpressured layer chromatography, the layer is developed repeatedly in the same direction, with each subsequent run progressing over a longer migration distance using a solvent of lower elution strength. Between runs, the solvent is completely removed from the chamber and the layer is dried under vacuum. The combined focusing effect of multiple development and gradient elution results in narrow bands, so that many components can be separated in a short migration distance. The 2D methods require two separated developments using two solvents of different selectivity. The test extract is spotted in one corner with reference standards on the two adjacent corners. The plate is developed in one direction, then rotated (90°) and developed in a second direction.

The 2DTLC is a very powerful technique and offers greater resolution than most other chromatographic techniques. In late January 2000, US corn was detained at the Port of Bar because of alleged contamination with >300 ng/g aflatoxin G_2 determined by HPLC . Results of the analysis were reported in newspapers and on television with comments that the United States was sending toxic corn to Montenegro. The HPLC and one-dimensional TLC chromatograms showed a huge amount of aflatoxin G_2. Simple 2 dimensional TLC demonstrated to the local laboratory that there was no G_2 in the corn. The corn was released and marketed immediately.

Analytical variability and errors of interpretation can be minimized by techniques such as 2DTLC and knowledge of potential false positives. For example: with chloroform-acetone as mobile phase dried bean extracts give a blue fluorescent spot at an R_f similar to that for AFB_1, and corn extract produces a spot between AFG_1 and AFG_2 that could be misidentified as aflatoxins. Similarly, a peanut butter spot and a milk spot make the chromatogram of these commodities difficult to quantify for aflatoxins. The use of mobile phases of different selectivity such as ether-methanol-water or 2 D TLC can confirm the findings.

Observing, Estimating and Quantitating

After developing, the plate is removed from the developing tank and the solvent is evaporated at room temperature in a chemical hood for a few minutes. The plate is then examined under ultraviolet (UV) light for fluorescent mycotoxins such as aflatoxins, ochratoxins, zearalenone and citrinin, or the plate is sprayed with a chemical reagent to form colored or fluorescent derivatives for mycotoxins such as deoxynivalenol, nivalenol, patulin, cyclopiazonic acid and the fumonisins. The fluorescence or color intensities of the presumptive toxin spots from the extract are compared with those of standard toxin spots visually or are measured with densitometer. A computerized densitometer can scan, record, and compute the amount of toxin in standards, and the extract spots and calculate the concentration of the toxin in the original extracted test portion.

Modern instrumental video techniques for TLC have been applied to the quantitative analysis of aflatoxins. The detection limits of the aflatoxins are in the low picogram range. Video densitometry is based on a charge-coupled device (CCD) camera housed in a view-box with an ultraviolet transilluminator, which causes the aflatoxins to emit a blue-green fluorescence. A sensitive video camera detects the fluorescence which is next to a digitizing board which processes the image on a personal computer. Special software has been developed for on-screen manipulation of image documents by selecting scan locations, data extraction, and integration. The main advantage of video technology is speed (a few seconds, compared with 20 min with slit-scanning densitometry) and its ability to provide an achievable image and chromatographic data.

MYCOTOXIN WORKSHOP ANALYTICAL PROCEDURE FOR AFLATOXINS

Thin layer chromatography (TLC) for the aflatoxins has received the most attention over the years; consequently it is the most refined and generally serves as a model for the other mycotoxins. The AOAC *Official Method 994.08* described below may be used to analyze corn, almonds, Brazil nuts, peanuts and pistachio nuts for aflatoxins. A recovery sample should be analyzed with each set of samples.

Apparatus

    a. High speed explosion proof blender

    b. Cleanup column – Mycosep 224 MFC column, Romer Labs® Inc., Washington, MO 63090.

    c. TLC plates – E. Merck Silica Gel 60, 20 x 20 plates

    d. Micro-syringe – 10 µL

    e. Spotting template

    f. TLC developing tank – Glass

    g. UV illumination cabinet – 365 nm

Reagents

    a. Solvents – acetonitrile, chloroform, acetone, 2-propanol, trifluoroacetic acid

b. Extraction solvent – Mix 900 mL acetonitrile with 100 mL water

c. Aflatoxin standard solutions – Prepare a standard solution of aflatoxins B_1, B_2, G_1 and G_2 in toluene-acetonitrile (9 + 1). The concentration is 0.5 µg aflatoxins B_1 and G_1/mL and 0.15 µg aflatoxins B_2 and G_2/mL.

d. Spotting solvent – toluene-acetonitrile (9 + 1)

e. TLC developing solutions

   1. Chloroform-acetone-water (90 + 10 + 1)

   2. Chloroform-acetone-2-propanol (85 + 10 + 5)

Extraction and Cleanup

a. Weigh a 50 g test portion into blender jar. Add 100 extraction solvent. Blend 2 minutes at high speed. Filter and collect the extract. Pipet 6 mL into 10 mL culture tube.

b. Hold cleanup column in one hand, and culture tube containing 6 mL extract in the other hand. Slowly push cleanup column into culture tube. Collect 4 mL into a 4 mL vial. Evaporate to dryness.

Spotting Samples on TLC Plates

a. Add 200 µL spotting solvent to each sample vial. Vortex to dissolve sample residue.

b. To prepare a plate for chromatography, draw a line 3 cm from the top to act as a stop for solvent and 1 cm from each side to prevent edge effects.

c. Starting from the left side, spot samples and standards according to the following template.

Thin Layer Chromatographic Spotting Template for Quantitation

Spots are 1 cm apart; 1ˢᵗ spot, 2 cm from bottom edge, and 2 cm from side of plate, spot only 2 test extracts on one plate

Spots:

· · · · · · · · · · · · · · · · ·

1 2 3 4 5 6 7 8 9 10 11 12 13 14 15 16 17

Step 1   Spot test extract (1) 2, 5, 10, 10 μL (spots 1-4)

Step 2   Spot standard 2, 5, 7, 10 μL (spots5-8).

                Standard: solution of Aflatoxin B_1 (0.5 ng/μL), B_2 (0.15 ng/μL), G_1 (0.5 ng/μL), and G_2 (0.15 ng/μL)

Step 3   Spot standard 2, 5, 7, 10 μL (spots 13-16).

                Standard: solution of Aflatoxin B_1 (0.5 ng/μL), B_2 (0.15 ng/μL), G_1 (0.5 ng/μL), and G_2 (0.15 ng/μL)

Step 4   Spot test extract (2) 2, 5, 10, 10 μL (spots 9-12)

Step 5   Spot standard 1 μL (spot 17)

Step 6   Add 5 μL standard on the 2nd 10 μL spot of test sample.

Plate development

a. Developing systems must be tailored to individual laboratories because laboratory conditions may be different, i.e. room temperature, humidity, spotting technique. To be on the safe side it is recommended that the analyst check the TLC development with standards before working with the test extract of interest. Development of the plates should be done in a fume hood regardless of which developing solvents are used.

b. Pour approximately 25 mL of developing solvent into the solvent trough in the tank. Insert the plate in the trough with the upper edge leaning against the wall of the tank, and seal with cover. Development to 17 cm stopline normally takes about 45 min.

c. Optimum development should give the complete separation of aflatoxins B_1, B_2, G_1 and G_2 in descending Rf order with AFB_1 approximately 5 to 8 cm from the origin. If the plate is left too long in the tank spots may become diffuse or remix at the solvent front.

d. After the plate has been developed, remove it from the tank and allow the solvent to evaporate to dryness. Residual solvent especially acetone affects the accuracy of quantitation.

Interpretation of the TLC Chromatogram

a. Examine the plate under long wave UV light in a UV viewing chamber. The four aflatoxin spots should be separated from each other in clearly defined spots and separated from non-aflatoxin fluorescent spots in the extract of the commodity being examined. The order of appearance of the spots is aflatoxin B_1, B_2, G_1 and G_2 from top to bottom.

b. Examine and compare the pattern of the spots from the sample extract containing internal standard for aflatoxin spots. R_f values of aflatoxins used as internal standards should be the same as, or only slightly different from, those of respective aflatoxin standard spots. R_f is defined as the distance the spot migrated divided by the distance the solvent migrated.

c. The fluorescent spots in the test extracts (presumptive aflatoxins) must have R_f values and color identical with those of the aflatoxin standard spots. In the sample extract containing standard, the presumptive aflatoxin spots and the internal standard spots are superimposed.

d. Compare fluorescent intensities of B_1 spots of the test extract with those of the standard spots and determine which standard spot matches the test extract.

e. If the intensity of the sample spot is between those of two standard spots then interpolate between the standard spots to determine the concentration value.

f. If spots of the smallest portion of the test extract are too intense to match the standards, dilute a portion of the sample extract to a proper volume with toluene-acetonitrile solution to make a more diluted solution of the test extract for respotting.

g. Compare fluorescent intensities of aflatoxins B_2, G_1 and G_2 in a similar manner.

h. If the aflatoxin plates show/contain interferences, spots are not well separated, or conditions are apparent which make the results doubtful, then the TLC should be repeated using a different solvent or TLC technique as described above.

Calculations

Calculate the concentration of aflatoxin B_1 in ng/g (= µg/kg, parts per billion, ppb) from the following formula:

$$ng/g = (S \times Y \times V)/(X \times W)$$

Where:

S = µL aflatoxin B_1 standard equal to unknown;

Y = concentration of B_1 standard (µg/mL);

V = µL of final dilution of test extract;

X = μL of test extract spotted giving fluorescent intensity equal to S (B_1 standard)

W = g of test sample corresponding to test extract used for TLC

Confirmation of Aflatoxin by Chemical Derivative Formation

To confirm the identity of the aflatoxins the following procedure is used.

Thin Layer Chromatographic Spotting Template for Confirmation

Spots are 1 cm apart; 1st spot, 2 cm from bottom edge, and 2 cm from side of plate, spot only 4 test extracts on one plate

Left side of plate

. . . .

1 2   3   4   5   6   7   8   9

Right side of plate

. . . .

10 11   12   13   14   15   16   17   18

Trifluoroacetic acid (TFA) procedure

a. Draw a line down the center of the TLC plate. On the left side of the plate spot two 10 μL portions of the test extract of each test sample (spot 1 and 2 for test extract of sample 1). Spot 5 μL standard solution (spot 9). Add 5 μL standard on top of one of the test extract spot (spot 2 for test extract of test sample 1).

b. On the right side of the plate spot two 10 μL portions of test extract of each test sample (spot 10 and 11 for test extract of sample 1). Spot 5 μL standard solution (spot 18). Add 5 μL standard on top of one of the test extract spot (spot 11 for test extract of test sample 1).

c. Cover right side of the plate with glass to avoid contamination of spots with TFA vapors.

d. Add 2 μL TFA to all spots on the left side of the plate with a disposable micropipette.

e. Place the plate in a dark area and let the TFA react for 5 minutes.

f. Place plate in oven at 45°C for 10 minutes.

g. Let plate cool and develop with chloroform-acetone-2-propanol (85 + 10 + 5).

h. After development, let solvent evaporate. Examine plate under long wave UV light. Unreacted aflatoxins appear near the top of right side of the plate without TFA. Blue fluorescent derivatives (B_{2a} and G_{2a}) appear at R_f about ¼ that of B_1 and G_1.

i. If the R_fs of B_{2a} and /or G_{2a} in derivatized test spot channel and the derivatized aflatoxins with the superimposed standards are the same as the R_fs of B_{2a} and G_{2a} in the standards only channel, then the identity of the aflatoxins in the test sample is confirmed.

13. TWO THIN LAYER CHROMATOGRAPHIC METHODS FOR THE SIMULTANEOUS ANALYSIS OF AFLATOXIN, ZEARALENONE AND DEOXYNIVALENOL AND FOR SELECTED TYPE A AND B TRICHOTHECENES.

John L. Richard, Jonne Hanneken, Keith Fleetwood, George Smiley, and Donna Houchins

Romer Labs, Inc.,
1301 Stylemaster Drive,
Union, MO 63084

Thin layer chromatography (TLC) is a very popular and widely used separation technique in analytical chemistry. It is a powerful screening tool but still offers a quantitative assessment of analytes in a wide variety of matrices (1). Currently new methods are continuously being developed. However, except for some new applications these methods are not often published in the literature (2). The simplicity of TLC, its low cost, fairly rapid procedures, need for limited technical expertise and facilities, along with the powerful nature of this test, make it a very desirable test to use.

Thin layer chromatography is the separation of compounds in a mixture, usually an extract of a solid matrix. Quantities of this mixture are placed in small localized areas (spotted) near one end of a plate (commonly glass) that is coated with a thin layer of an adsorbent matrix such as silica gel. Standards of the analyte of interest are spotted on the plate also. Separation occurs when the end of the plate nearest the spotted mixture is placed in a solvent system in the bottom of a closed vessel (glass tank or jar). The solvent is allowed to migrate through the adsorbent matrix and moves toward the top of the plate. As this occurs the mixture of compounds separate based on their interactions between the solvent system (mobile phase) and the matrix (stationary phase). Because these properties differ for the various compounds in a mixture, they will migrate on the plate at different rates from their starting point. If the analyte of interest (mycotoxin in this case) is present in the mixture it will be identified by comparison with the standards.

In this report we describe two methods for TLC analysis of several important mycotoxins that occur in grains, and other commodities and products worldwide. One of the methods is for the simultaneous analysis of aflatoxins, zearalenone and deoxynivalenol (DON) (3-toxin test) and the other method is for the simultaneous analysis of several type A and B trichothecenes.

METHODS AND MATERIALS

Equipment

(a) Balance, 400g, Model Scout, Ohaus Corp. Florham Park, NJ

(b) Timer, Electronic Clock/Double Timer, West Bend Co., West Bend, WI

(c) Blender, Osterizer 14 speed, blender bases and blade assemblies for half-pint jars, Sunbeam-Oster Household Products, Schaumburg, IL or Gyrotary shaker, fitted with 250 mL Erlenmeyer flask holders, Eberbach Corp., Ann Arbor, MI

(d) Pipettors

Finnpipette 1-5mL, Labsystems, Helsinki, Finland

Finnpipette 200-1000 µL, Labsystems, Helsinki, Finlan

(e) Water bath, VWR Scientific Products, Model 1230, Sheldon Mfg., Inc., Cornelius, OR , or Romer EVap', Romer Labs®, Union, MO

(f) Mixer, Vortex Genie 2, Model G-560, Scientific Industries, Inc., Bohemia, NY

(g) Ultraviolet light, Long Wave, Blak-Ray Lamp, Ultraviolet Products, Inc., San Gabriel, CA or Viewing box, C-10 Chromato-vue, VWR, Pittsburg, PA

(h) Hot plate, Thermodyne Type 2200, Barnstead/Thermodyne, Dubuque, IA

(i) Thin layer chromotogram Autospotter*, Romer Labs, Inc., Union, MO

(j) 250 mL Erlenmeyer flasks fitted with #6 Neoprene stoppers or half pint blender jars

(k) 40 mL conical tubes with #5 Neoprene stoppers

(l) Test tube rack for 40 mL conical tubes

(m) Test tube rack for 16 x125 and 15 x 85 culture tubes

(n) 250 mL sample jars with lids

(o) Funnels, utility, polypropylene, 108 mm

(p) 1,2, 10 and 20 mL volumetric pipettes

(q) 25 µL syringe, Unimetrics, Fuson, CA

(r) 250 mL Erlenmeyer flasks with sidearm and fitted with #6 rubber stopper

(s) 18 gauge syringe needles, Fisher Scientific, Fairlawn, NJ

(t) 100 µL syringes for Autospotter*

(u) Thin layer plate developing tank (12 in. long x 6 in. high x 4 in. wide) with glass plate lid

(v) Tongs

(w) Reagent spray bottle

(x) 100 mL graduated cylinders

Consumables

(a) 16 x 125 borosilicate culture tubes

(b) 15 x 85 borosilicate culture tubes

(c) 12 x 75 cuvettes with caps

(d) Solid phase cleanup columns, MultiSep 216, Romer Labs, Inc., Union, MO

(e) Solid phase cleanup columns, MycoSep 227, Romer Labs, Inc., Union, MO

(f) Solid phase cleanup columns, MycoSep 226, Romer Labs, Inc., Union, MO

(g) Pipette tips for 1-5 mL pipettors,

(h) Pipette tips for 200-1000 mL pipettors

(i) Filter paper, qualitative

(j) Thin layer plates, silica gel HL, 10 x 10 cm, Analtech, Inc., Newark, DE

(k) Thin layer plates, reverse phase C_{18} silica gel, 10 x 10 cm, Whatman, Inc., Clifton, NJ

Reagents

(a) Acetic acid, glacial, Certified ACS Plus, Fisher Scientific, Fairlawn, NJ

(b) Acetone, Certified ACS, Fisher Scientific, Fairlawn, NJ

(c) Acetonitrile, HPLC Grade, Fisher Scientific, Fairlawn, NJ

(d) Aluminum chloride, Hexhydrated, 99%, Aldrich Chemical Co., Milwaukee, WI

(e) Methanol, HPLC Grade, Fisher Scientific, Fairlawn, NJ

(f) Sulfuric acid, Certified ACS Plus, Fisher Scientific, Fairlawn, NJ

(g) Toluene, HPLC Solvent, J.T. Baker, Phillipsburg, NJ

(h) Water, deionized

Standards

(a) 5 mL 3-toxin spiking standard: 1.25 µg/ml aflatoxin B1 and 62.5 µg/ml of each DON and zearalenone in acetonitrile.

(b) 5 mL 3 toxin working standard: 0.1 µg/mL aflatoxin B1 and 10 µg/mL of each DON and zearalenone in 95: 5 (v/v) toluene: acetonitrile.

(c) 5 mL trichothecene spiking standard: 100 µg/mL T-2 toxin (T-2), HT-2 Toxin, (HT-2), diacetoxyscirpenol (DAS), neosolaniol (NEOS), deoxynivalenol (DON), nivalenol (NIV), and fusarenon X (FX).

(d) 5 mL Type B trichothecene working standard: 10 µL/mL DON, NIV and FX in 2:1 (v/v) acetone: methanol.

(e) 5 mL T-2/DON silage spotting standard: 10 µg/mL each of T-2, DON, 3- acetyl DON and 15-acetyl DON in 2:1 (v/v) acetone:methanol.

(f) 5 mL Type A trichothecene working standard: 10 µg/mL T-2, HT-2, DAS and NEOS in 97:3 (v/v) toluene :acetonitrile.

METHOD FOR AFLATOXIN, DON AND ZEARALENONE (3-TOXIN TEST)

Extraction

Place 25 g of ground sample in a 250 mL Erlenmeyer flask or a half-pint, blender jar if blending is the extraction method. To the flask or jar, add 100 ml of 84:16 (v/v) acetonitrile:water. If a flask is used, shake for 1 hr on gyrotary shaker; if a blender jar is used, blend for 3 min. Filter extracts through suitable qualitative filter paper and collect filtrate (sample extract).

Purification

Prepare a solvent blank fortification or "neat spike" (analyte added to the extraction solvent) using the 3-toxin spiking standard by adding 45 µL of the standard to 7 mL of 84:16 acetonitrile:water in a 15 x 85 mm culture tube. Mix well and **treat this tube in the same manner as the sample through the remainder of method**. (*Note*: 100% recovery results in 4 ng of aflatoxin B_1 and 200 ng of each DON and zearalenone on the TLC plate). Place 7 ml of sample extract into a 15 x 85 mm culture tube. Add 25 mL of glacial acetic acid. Mix well and push approximately 2.5 mL of sample or neat spike through a MycoSep® #226 column. Quantitatively transfer (using a pipette) 2 mL of purified extract to a 12 x 75 mm cuvette. Evaporate to dryness under vacuum in a 60 °C water bath or use the Romer® Evap system.

TLC Determination

Dissolve residue remaining after evaporation in each tube in 400 µL of 95:5 (v/v) toluene:acetonitrile, stopper and place on vortex mixer for 30 seconds. With the syringes for the Autospotter®, pull up (draw) 100 µL of each sample and spike and place onto the Autospotter®. In a similar manner place syringes with 5, 10 and 20 µL of the 3-toxin working standard onto the Autospotter®. Place a Silica Gel TLC plate onto the Autospotter®, position the needles lightly against the plate gel surface and start the Autospotter®. After spotting, ensure that the spots are dry and then place the plate, with edge near the spots down, in a glass developing tank or jar in which there is 1:1 (v/v) toluene:acetone developing solution in the bottom. Cover the glass tank or jar and let stand until the solvent front reaches within 1 cm of the top edge of the plate (approx. 20 min). Remove the plate from the tank and allow to air dry in a ventilated hood.

View plate under long wave UV light for aflatoxin B_1 and zearalenone and compare with standards applied to the plate. Then spray the TLC plate with 15% aluminum chloride in methanol and let the plate air dry in a ventilated hood. View the plate under long wave UV light for zearalenone again and compare with standards applied to the plate. Then heat the plate at 150 °C until the DON spots are visible. Compare any spots with DON standard spots.

METHOD FOR TRICHOTHECENES

Extraction

Place 25 g of ground sample into a 250 mL Erlenmeyer flask or a half-pint, blender jar if a blender is used for extraction. Add 100 mL of 84:16 v/v acetonitrile:water to flask or jar and stopper or seal with blender assembly, respectively. If flask is used, shake for 1.5 hours on the gyrotary shaker; if blender is used, blend for 3 min. Filter extracts into sample jar through suitable qualitative filter paper.

Aqueous samples: Combine 4.0 mL of aqueous sample with 21.0 mL of acetonitrile in a 40 mL conical tube, close with neoprene stopper and mix on vortex mixer for 1 min.

Cleanup Preparation

Prepare for sample cleanup by attaching a MultiSep® 216 column to a vacuum apparatus and fill column with 84:16 (v/v) acetonitrile:water. Adjust flow rate to approximately 1 mL/ min. Discard the wash.

Prepare a matrix or neat spike by adding 25 µL of Type A and B trichothecene spiking standard to 10 mL of filtered extract or 10 ml of

84:16 (v/v) acetonitrile:water, respectively, into a 15 x 85 mm culture tube. Mix well and **treat this tube in the same manner as the sample through the remainder of the method.** (*Note*: 100% recovery results in 200 ng of each trichothecene on the TLC plate).

Aqueous solutions: For aqueous solutions add 16 µL of Type A and B trichothecene standards to 10 mL of 84:16 (v/v) acetonitrile:water and mix well.

Place a clean 16 x 125 mm culture tube inside the side arm flask on the vacuum apparatus for collection of the sample and the wash.

Purification

After pipetting 10 mL of filtered extract into a 15 x 85 mm culture tube, push just over 4 mL of sample or spike through a MycoSep® 227 column. Transfer 4 mL of purified extract to a washed MultiSep® 216 column and collect in a 16 x 125 culture tube. The flow rate should be approximately 1 mL/ min.

Aqueous solutions: For aqueous solutions, push 10 mL of solution completely through the #227 column. Transfer 6.25 mL of purified extract to a washed MultiSep® 216 column and collect the eluate (the portion that passes through the column).

When the solution reaches the top of the packing, add two, 6 mL increments of 84:16 (v/v) acetonitrile: water to the cleanup column and continue to collect until there is no more eluate. Transfer the purified solution from the 16 x 125 mm culture tube to a 40 mL conical tube. Rinse the culture tube with 3 mL of acetonitrile and add to the 40 mL tube. Evaporate the solution to dryness under vacuum in a 60 °C water bath.

TLC DETERMINATION

For Type B Trichothecenes

Dissolve the residue in the conical tube in 400 µL of 2:1 (v/v) acetone: methanol, stopper and mix on vortex mixer for 30 sec. With the syringes for the Autospotter˙, pull up 80 µL of each sample and spike and place onto the Autospotter˙. In a similar manner place syringes containing 10, 20 and 30 µL of the Type B Trichothecenes working standard and 40 µL of T-2/DON silage spotting standard onto the Autospotter˙. Place a Silica Gel TLC plate onto the Autospotter˙, position the needles lightly against the plate gel surface and start the Autospotter˙. After spotting, ensure that the spots are dry and then place the plate, with edge near the spots down, in a developing tank into 1:2 v/v toluene:acetone developing solvent. Cover the tank and let stand until the solvent front reaches within 1 cm of the top edge of the plate (approx. 15 min). Remove the plate from the tank and allow to air dry in a ventilated hood. Spray the plate with 15% aluminum chloride in methanol and allow the plate to dry in a ventilated hood. Heat the plate at 150 °C on a hot plate until standard spots are fully visible under long wave UV light. Remove the plate from the hotplate and view plate under long wave UV light and compare with standards applied to the plate.

For Type A Trichothecenes

Evaporate the remaining spotting solution in the 40 mL tube to dryness under vacuum in a 60 °C water bath. Dissolve the residue in 320 µL 97:3 (v/v) toluene:acetonitrile, stopper and mix on vortex mixer for 30 sec. With the syringes for the Autospotter˙, pull up 80 µL of each sample and sample fortified with standard spike and place onto the Autospotter˙. In a similar manner place syringes containing 10, 20, 30 and 40 µL of the Type A Trichothecene working standard onto the Autospotter˙. Place a reverse phase C_{18} silica gel TLC plate onto the Autospotter˙, position the needles lightly against the plate gel surface and start the Autospotter˙. After spotting, ensure that the spots are dry and then place the plate, with edge near the spots down, in a developing tank into 25:15:1 (v/v/v) methanol:water:acetic acid developing solution. Cover the tank and let stand until the solvent front reaches within 1 cm of the top edge of the plate (approx. 45 min). Remove the plate from the tank and allow to air dry in a ventilated hood. Spray the plate with 10% sulfuric acid in methanol and allow the plate to dry in a ventilated hood. Heat the plate at 150 °C on a hot plate until standard spots are fully visible under long wave UV light. (*Note:* Do Not Scorch Plate). Remove the plate from the hotplate and view plate under long wave UV light and compare with standards applied to the plate.

RESULTS AND DISCUSSION

In the 3-toxin method above, aflatoxin B_1 will fluoresce a bright blue under longwave UV light with an R_f (position of the analyte relative to the location of the solvent front) of approximately 0.5. Zearalenone will appear blue after spraying with the aluminum chloride spray reagent and will have an R_f of approximately 0.7. After heating the plate DON spots will appear blue under UV light at an R_f of approximately 0.3. Quantification can be achieved by visual comparison with standard spots on the plate. For aflatoxin spiked at a level to theoretically provide 0.5 ng on the plate (N = 9), 100% recovery was attained and the limit of detection was 4 µg/kg (ppb). For zearalenone and deoxynivalenol spiked at a level to theoretically provide 50 ng on the plate (N = 9), 86 to 71% recovery was attained, respectively, and the limit of detection was 400 ppb. With the latter two mycotoxins, if the spiking was at a level to theoretically provide 200 ng on the plate, the recoveries for both mycotoxins was approximately 90%.

For the TLC plates for Type B trichothecenes, after spraying the plates with aluminum chloride spray reagent and heating as described, 3-acetyl DON will appear blue with an R_f of approximately 0.85, 15-acetyl DON will appear blue with an R_f of approximately 0.8, fusarenon X will appear blue with an R_f of approximately 0.7, DON will appear blue with an R_f of approximately 0.5 and nivalenol will appear blue with an R_f of approximately 0.2. Again quantification can be achieved by visual comparison with standard spots on the plate.

For the TLC plates for Type A trichothecenes, after spraying the plates with the sulfuric acid spray reagent and heating as described, neosolaniol will appear blue with an R_f of approximately 0.6, diacetoxyscirpenol will appear light purple with an R_f of approximately 0.5, HT-2 toxin will appear blue with an R_f of approximately 0.4 and T-2 toxin will appear blue with an R_f of approximately 0.3.

The thin layer chromatography methods employed above have been used for several years in the Romer® Labs, Inc. analytical services laboratory and, where applicable, have approximated values achieved using HPLC methods and quantitation. For the trichothecenes method, the recoveries of type A trichothecenes averaged 85.2% (N=5; range, 80-95%) and recoveries for Type B trichothecenes averaged 85.6% (N=20; range, 77.5-92.5). Detection limits in mg/g (ppm) for the trichothecenes were as follows: 3-acetyl-DON = 0.1; 15-acetyl-DON = 0.1; fusarenon-X = 0.5; deoxynivalenol = 0.1; nivalenol = 0.5; neosolaniol = 0.1; diacetoryscirpenol = 0.3; HT-2 toxin = 0.1; and T-2 toxin – 0.1. Although, quantitative assessment is achieved presently using visual means by comparison with standards, we recommend that if more exact quantitation is desired that densitometry be employed. Densitometry would have to be suitable for both fluorescent and colored spots on the thin layer chromatogram.

To achieve better quantitation using the visual method of comparison with standards applied to the plates, it is recommended that the Autospotter* described be used. This achieves spots that are kept very small and all spots on a plate are of a similar size so that comparisons are more accurate. The remaining materials for TLC are very inexpensive and can be conducted with limited space and equipment. A ventilated hood is recommended to limit technician exposure to solvents and reagents used in the process.

Presently, all of the materials to conduct the methods reported here can be acquired through the sources listed. However, for convenience to our clients, Romer Labs˚, Inc. has put together a package of all consumable materials, (except solvents which can be obtained locally), necessary for the analysis of 50 samples for either of the methods described. These are available from Romer Labs˚, Inc. as a TLC Supply Package for either the 3-toxin method for 50 samples or the trichothecene A and B method for

50 samples. All other equipment and supplies described for the method are available from the sources listed or from Romer Labs', Inc. as well.

Thin layer chromatography is useful for the direct quantitation of mycotoxins as described, but is often used for confirmation of mycotoxin analysis by other means such as HPLC or gas chromatography.

SUMMARY

Two thin layer chromatographic methods are described for the separation and identification of eleven different mycotoxins. One of the methods is for the simultaneous analysis of aflatoxins, zearalenone and deoxynivalenol. The other method is for 3-acetyl- deoxynivalenol, 15-acetyl deoxynivalenol, deoxynivalenol, fusarenon-X, nivalenol, neosolaniol, diacetoxyscirpenol, HT-2 toxin and T-2 toxin. Both of these methods are inexpensive and quantitative assessments can be made with adequate standards applied to the thin layer chromatographic plate. All materials for the analysis can be obtained easily and the tests can be conducted where space and equipment is limited.

REFERENCES

(1)   Touchstone, J.C., (1992) *Practice of Thin Layer Chromatography*, 3rd Edition, John Wiley and Sons, Inc., New York, NY 377

(2)   Trucksess, M.W., (2001) in de Koe, WJ, Samson, RA, van Egmond, HP, Gilbert, J, Sabino, M, (eds), *Mycotoxins and Phycotoxins in Perspective at the Turn of the Millennium*, W.J. de Koe, Hazekamp 2, Wageningen, The Netherlands, 29..

14. THE SIMULTANEOUS ANALYSIS OF AFLATOXINS, DEOXYNIVALENOL, FUMONISINS, OCHRATOXIN A AND ZEARALENONE USING MIXED-BED IMMUNOAFFINITY COLUMN AND HPLC/UV/FLUORESCENCE WITH POST-COLUMN DERIVATIZATION

Darsa Siantar[1]*, Maria Ofitserova[2], Sareeta Nerkar[2], Michael Pickering[2], Maria Cardozo[1], George Peterson[1] and Sumer Dugar[1]

[1]U.S. Department of the Treasury
Alcohol and Tobacco Tax and Trade Bureau (TTB)
490 North Wiget Lane,
Walnut Creek, CA 94598, USA

[2]Pickering Laboratories, Inc.
1280 Space Park Way;
Mountain View, CA 94043, USA

* corresponding author

The co-occurrence of multiple mycotoxins in an economically important agricultural product has been shown to be a common phenomenon. Combinations of two or more mycotoxins have been isolated from corn (1) and spices (2). Some agricultural products such as barley and grapes, may be used for purposes other than direct food consumption. The occurrence of multiple potentially toxigenic fungal species has been demonstrated in barley (3, 4). Malt beverages could, therefore, potentially contain different mycotoxin combinations from fungi-infected grains used in their production. Ochratoxin A (OTA) is the predominant naturally-occurring mycotoxin in grape wines and numerous incidences of its occurrence have been reported (5-11). In addition to OTA, various fruit- and grain-based wines or beers could simultaneously be contaminated by the aflatoxins, deoxynivalenol (DON) among other trichothecenes such as T-2 and HT-2, fumonisins, and zearalenone (ZEA). The likely occurrence of two zearalenone metabolites, α- or β-zearalenol (the predominant metabolite), transmitted into beers from contaminated grains during brewing had also been reported (12). From a public health perspective, it is important that a capability to rapidly determine the presence of all toxins that may exist in a food product is developed. The need to have such methodologies readily available is even more so in rural communities of developing countries where the potential for exposure to multiple mycotoxins is real. For the methodology to be universally applicable, it should be able to detect the presence of trace-levels of mycotoxins in several commodities, including alcoholic beverages.

A prototype 5-toxin (AflatoxinDONFumonisinOchraZea) immunoaffinity column has been developed by Vicam (Watertown, MA) for the purpose of this collaborative project. It is a product of the consolidation or capacity enhancement of the earlier 2-toxin (AfaltoxinOchra or AflaOchra HPLC™), 3-toxin (AflatoxinOchraZea or AOZ HPLC™), 3-toxin (AflatoxinDONFumonisin) and 4-toxin (AflatoxinDONZeaOchra) antibody-based SPE columns. These customized IA columns loaded with multiple antibody-bound resins use antibody-antigen retention mechanisms to bind with different mycotoxins. Consequently, only one solid-phase extraction (SPE) is required to isolate, and quantify up to 5 types of mycotoxin families. The single multi-toxin extract may be used to determine the various toxins using a fluorometer or any other appropriate instrumentation.

Previously developed LC methods have combined up to three single-mycotoxin families (aflatoxins, ochratoxin A and zearalenone) into a single run (13, 14). The Alcohol and Tobacco Tax and Trade Bureau

(TTB), in collaboration with Pickering Laboratories, Vicam, R-Biopharm Rhône, Waters, Varian and Food and Drug Administration (FDA) set out to develop analytical system combinations that would use the prototype 3-toxin (AflatoxinDONFumonisin), 4-toxin (AflatoxinDONZeaOchra), and 5-toxin (AflatoxinDONFumonisinOchraZea) IA columns for sample clean-up and/or enrichment and HPLC for separation and detection of trace levels of multiple mycotoxins in alcoholic beverages.

The new separation methods use HPLC with an in-line photochemical reactor, post-column derivatization, and UV and fluorescence detection, to detect up to five different mycotoxin families in groups of 3-, 4- or 5-toxin. These LC methods could be used to monitor the existence of possible different mycotoxin combinations in alcoholic beverages, with sensitivities well below the current regulatory limits. This paper only describes the use of the 5-toxin (AflatoxinDONFumonisinOchraZea) LC method to evaluate the efficacy of SPE column performance in extraction, isolation and enrichment of naturally-occurring mycotoxins in alcoholic beverages (beer and rice wine).

MATERIALS AND METHODS

Reagents and Chemicals

o-Phthalaldehyde (OPA), OPA diluent (5.4% potassium borate in water), Thiofluor™ (N,N-Dimethyl-2-mercaptoethylamine hydrochloride), and phosphoric buffer solution (P/N 1700-1108) were provided by Pickering Laboratories (Mountain View, CA). 30% (w/v) Brij° 35 solution was obtained from Sigma (St. Louis, MO). Acetonitrile and methanol (both HPLC grade) were obtained from Fisher Scientific (Pittsburgh, PA). Deionized water was produced in the laboratory using a Millipore Milli-Q system (Bedford, MA). Aflatoxins B_1, B_2, G_1 & G_2, deoxynivalenol, ochratoxin A, and zearalenone standards in solution were obtained from Supelco (Bellefonte, PA). Fumonisins B_1, B_2 & B_3 were donated by PROMEC, Medical Research Council (Tygerberg, South Africa). The prototype 5-toxin (AflatoxinDONFumonisinOchraZea) monoclonal antibody-based columns and phosphate buffered saline (PBS) 10x concentrate solution were donated by Vicam (Watertown, MA). SurfaSil™ siliconizing fluid for surface treatment of in-house laboratory glassware was obtained from Pierce Biotechnology (Rockford, IL).

Reagent Preparation

OPA reagent preparation: 950-mL OPA diluent was sparged with helium (inert gas) for 15 min in a 1L bottle. 300-mg OPA was dissolved in 10 mL methanol and 2-g of Thiofluor™ was then added to the OPA in methanol solution. The inert gas was turned off and the OPA and Thiofluor™ mixture was added to the sparged diluent. 3-mL of 30% (w/v) Brij° 35 solution was also added to the mixture and the contents thoroughly mixed. The OPA reagent was kept under inert gas to minimize oxidation.

Multi-toxin stock and working standard solution preparation: Known amounts of the five mycotoxin families were transferred into a silanized borosilicate glass volumetric flask The solvent was dried and the residue reconstituted in methanol+water (1+1) to the mark, to produce a primary multi-toxin stock standard solution. The primary multi-toxin stock standard solution was used to spike alcoholic beverage samples at various levels and prepare secondary mixed-toxin working standard calibration solutions at five different concentrations. The 5 secondary mixed-toxin working standard solutions for calibration purposes were freshly prepared for each experiment. The primary multi-toxin stock standard solution was stored in the freezer (~ -20 °C) when not in use.

Equipment and Analytical Conditions

The HPLC system consisted of an Agilent 1100 Series quaternary pump and injection system, including a standard autosampler, fluorescence detector (FLD) and diode-array detector (DAD) from Agilent Technologies (Palo Alto, CA). The complete system apparatus comprised of several instruments that were assembled in series (HPLC – DAD or commonly known as ultra-violet (UV) detector – photochemical reactor – post-column derivatizer – fluorescence detector – waste). The analytical column was a MycoTox™ C_{18}, 4.6 x 250 mm, 5 µm particle size coupled to a 5-μm guard column both from Pickering Laboratories (Mountain View, CA). The mobile phase consisted of combinations of three reagents (Table 14.1). The flow rate was 1 mL/min with column temperature set at 40 °C and sample injection volume of 30 μL. The equilibration time was 10 minutes.

Table 14.1. HPLC gradient settings

Time	Phosphoric buffer (P/N 1700-1108), %	Methanol, %	Acetontrile, $
0.0	85	0	15
5.0	85	0	15
5.1	57	28	15
20.0	57	28	15
23.0	40	60	0
40.0	40	60	0
50.0	0	100	0
60.0	0	100	0

The Photochemical Reactor for Enhanced Detection ("PHRED"™) unit from Aura Industries (New York, NY) was equipped with a 254 nm low pressure Hg lamp and PTFE (poly-tetrafluoro-ethylene) knitted reactor coils. The 254-nm UV light was able to perform continuous photolytic derivatization to enhance the sensitivity and/or selectivity of fluorescence detection response for aflatoxins in particular. The PCX5200 post-column derivatization system was equipped with a Control Software from Pickering Laboratories (Mountain View, CA). The reactor volume and temperature were set at 1.4 mL and 65 °C, respectively. The derivatizing reagent was OPA. The flow rate was set at 0.3 mL/min. The post-column pump program was activated using the PCX5200 Control Software by turning the pump on at 23.0 minutes, then off at 34.5 minutes and on again at 43.5 minutes and finally off at 60.0 minutes. The HPLC gradient, detector settings and programming are as shown in Tables 14.1-14.3. All gradient and wavelength changes were programmed with the Agilent ChemStation software that was used for data acquisition and management.

Table 14.2. Mycotoxin specific wavelength settings of ultra-violent and fluorescence detectors

Analyte	Derivatization	Detection	Wavelength
DON	None	Ultra-violet	$\lambda = 218$ nm
Aflatoxins	Photolytic (PHRED™)	Fluorescence	λex = 365 nm λem = 455 nm
Fumonisins	Post-Column (OPA)	Fluorescence	λex = 330 nm λem = 465 nm
Ochratoxin A	None	Fluorescence	λex = 335 nm λem = 455 nm
Zearalenone	None	Fluorescence	λex = 275 nm λem = 455 nm

Table 14.3. Wavelength programming on the fluorescence detector

Time	λex	λem
0.0	365	455
26.0	365	455
26.1	330	465
36.0	330	465
36.1	335	455
41.0	335	455
41.1	275	455
44.0	275	455
44.1	330	465
60.0	330	465

Sample Preparation and SPE Column Clean-up Protocols

The Visiprep' 24-port SPE vacuum manifold and Visidry' drying attachment from Supelco (Bellefonte, PA) were used for sample preparations. About 100 mL alcoholic beverage was degassed for 15 minutes using a 250 mL aspirator bottle placed in a sonicator. 25-mL of the degassed alcoholic beverage was transferred to a 25 mL volumetric flask and spiked with 182 µL of the primary multi-toxin stock standard solution to obtain 163 µg/L DON, 0.485 µg/L G_2, 1.542 µg/L G_1, 0.492 µg/L B_2, 1.527 µg/L B_1, 18 µg/L FB_1, 18 µg/L FB_2, 20 µg/L FB_3, 0.80 µg/L OTA and 20 µg/L ZEA concentrations. A 5-mL aliquot of the spiked alcoholic beverage was placed in a salinized borosilicate culture tube, evaporated at room temperature to dryness by blowing "zero grade air" (nearly moisture and hydrocarbon free air) and reconstituted to the original volume in one-tenth diluted PBS solution. The entire contents were passed through the customized 5-toxin antibody-based SPE column (~1 drop/sec). A control 5 mL aliquot consisting of only diluted PBS solution spiked with multi-toxin standards was also passed through a customized 5-toxin antibody-based SPE column (~1 drop/sec). The column was washed with 4 mL deionized water (~2 drops/sec) and the mycotoxins eluted with 3 mL methanol (~1 drop/sec). The methanolic eluate was collected in a silanized borosilicate culture tube, air dried with "zero grade air" and reconstituted in 5 mL methanol. The final 5 mL mixed-toxin sample was vortexed (Vortex-Genie™, Scientific Industries, Bohemia, NY) and 30 μL extract injected onto HPLC.

Table 14.4 shows multi-toxin primary (stock) standard concentrations, alcoholic beverage spiking levels expressed as mass of toxin in ng loaded onto the mixed-bed IA column, and the 5-point standard calibration ranges utilized to generate the spike recovery data.

Table 14.4. Deoxynivalenol (DON), aflatoxins (G_1, G_2, B_1, B_2), fumonisins (FB_1, FB_2, FB_3), ochratoxin A (OTA) and zearalenone (ZEA) stock standard concentrations, mass loaded onto the column and 5-point standard calibration concentration ranges with correlation coefficients ≥ 0.999

	Multi-toxin stock standard concentrations (μg/mL)	Toxin masses loaded onto column (ng)	5-Point standard calibration concentration ranges (μg/L)
DON	22.392	814	56 - 896
G_2	0.067	2.4	0.2 - 2.7
G_1	0.212	7.7	0.5 - 8.5
B_2	0.068	2.5	0.2 - 2.7
B_1	0.210	7.6	0.5 - 8.4
FB_1	2.450	89	6 - 98
FB_2	2.500	91	6.2 - 100
FB_3	2.800	102	7 - 112
OTA	0.110	4.0	0.6 - 4.4
ZEA	2.762	100	14 - 110

Multi-toxin Detection Strategy

The ultra-violet detector and photochemical reactor were strategically placed in series before the post-column derivatizer hardware for the simultaneous UV detection of DON and photolytic derivatization of aflatoxins. The method allowed for fluorescent detection of, the aflatoxins after undergoing photolytic sensitization in a photochemical reactor, the fumonisins after OPA derivatization and, the ochratoxin A and zearalenone which fluoresce naturally. Fumonisins have a primary amine group that is derivatized post-column with OPA and a mercaptan to form highly-fluorescent adducts, 1-alkyl-2-thioalkyl-subsituted isoindoles exhibiting optimal excitation at 330 nm and maximal emission at 465 nm. The OPA reagent flow was started after the aflatoxins elution and stopped after the fumonisin B_1 peak elution. Sufficient delay time was allocated to flush the OPA from the tubing prior to the ochratoxin A and zearalenone elution. Then, the OPA reagent flow was turned back on during the fumonisins B_3 and B_2 elution. The fluorescence detector was time-programmed to change excitation and emission wavelengths for multi-toxin response optimization.

RESULTS AND DISCUSSION

A novel immunoaffinity/HPLC method has been developed to simultaneously analyze DON, ochratoxin A, zearalenone, aflatoxins, and fumonisins with post-column photochemical and *o*-Phthalaldehyde (OPA) derivatizations. The immunoaffinity sample preparation technique combined five single-antibody IA media into one SPE cartridge. In this study, all 5-point multi-toxin standard calibration curves generated had linear regression correlation coefficients ≥ 0.999.

Figure 14.1 shows a chromatogram of an alcoholic beverage spiked at a low multi-toxin level. The chromatogram demonstrates the efficacy of the 5-toxin IA column in cleaning up sample-matrix effects since the baseline was unaffected by any significant matrix interferences even at trace-level multi-toxin fortification

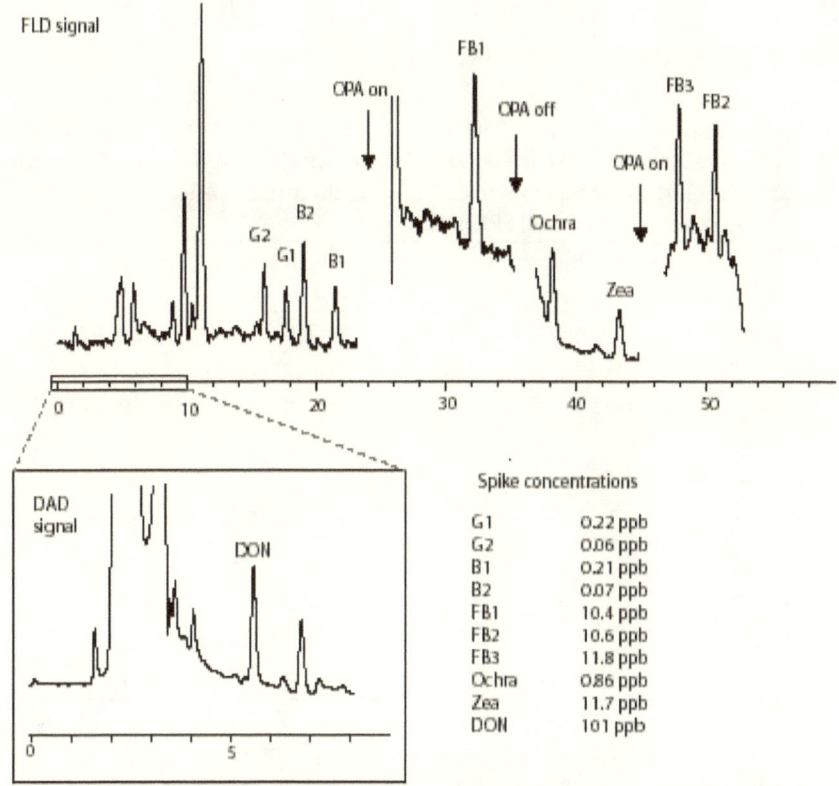

FLD denotes fluorescence detector, DAD represents diode-array detector,
and ppb equals to µg/L. Horizontal axis in min.

**Figure 14.1. Representative deoxynivalenol (DON), aflatoxins (G1, G2, B1,
B2), fumonisins (FB1, FB2, FB3), ochratoxin A (OTA) and zearalenone (ZEA)
HPLC (AflatoxinDONFumonisinOchraZea) antibody-based SPE column.**

The multi-toxin recoveries in spiked PBS control and alcoholic beverage
samples (beer and rice wine) ranged from 71 to 112 % with $RSD_r \approx$
10% (Table 14.5). The acceptable multi-toxin spike recovery range
demonstrated the prototype mixed-bed IA column's ability to effectively
and selectively bind with the targeted mycotoxins. The method detection
limits were 20 µg/L for deoxynivalenol, 0.02 µg/L for aflatoxin G_2, 0.08
µg/L for aflatoxin G_1, 0.01 µg/L for aflatoxin B_2, 0.02 µg/L for aflatoxin
B_1, 3 µg/L for fumonisin B_1, 3 µg/L for fumonisin B_2, 8 µg/L for
fumonisin B_3, 0.14 µg/L for ochratoxin A, and 9 µg/L for zearalenone.

The alcoholic beverage samples used in this study did not contain naturally-occurring mycotoxins.

Table 14.5. Mean deoxynivalenol (DON), aflatoxins (G_1, G_2, B_1, B_2), fumonisins (FB_1, FB_2, FB_3), ochratoxin A (OTA) and zearalenone (ZEA) recoveries (% ± SD) from spiked[a] dilute PBS, light beer and rice wine using the 5-toxin (AflatoxinDONFumo nisinOchraZea) antibody-based SPE column

	Spiked PBS (n = 3)	Spiked light beer (n = 3)	Spiked rice wine[#] (n = 3)
DON	105 ± 5	94 ± 4	95 ± 3
G_2	79 ± 8	75 ± 12	94 ± 3
G_1	71 ± 9	73 ± 13	91 ± 4
B_2	112 ± 7	88 ± 8	99 ± 2
B_1	107 ± 6	93 ± 9	96 ± 2
FB_1	83 ± 8	77 ± 3	84 ± 13
FB_2	100 ± 12	99 ± 8	83 ± 4
FB_3	90 ± 11	93 ± 2	84 ± 8
OTA	96 ± 2	102 ± 7	89 ± 4
ZEA	88 ± 3	101 ± 7	111 ± 6

[a]Samples spiked at 163 µg/L DON, 0.49 µg/L G_2, 1.54 µg/L G_1, 0.49 µg/L B_2, 1.53 µg/L B_1, 18 µg/L FB_1, 18 µg/L FB_2, 20 µg/L FB_3, 0.80 µg/L OTA and 20 µg/L ZEA
[#]Cloudy rice wine with cooked particulate matter.

Aside from increasing the injection volume (> 30 µL aliquot) of mixed-toxin sample extract onto HPLC, the mycotoxin pre-concentration and enrichment steps during the sample preparation could also be used to increase the overall method sensitivity to allow accurate trace level quantification of naturally-occurring mycotoxins in alcoholic beverages. Larger sample volumes (> 5 ml spiked aliquots) could be added into the IA column if mycotoxin pre-concentration was desired. During the enrichment step, if 5 mL spiked sample aliquot was passed through the column and the eluate containing mycotoxins was dried and reconstituted in 1 mL methanol, then the original multi-toxin concentrations were enriched by 5 times. Hence, when analyzing neat alcoholic beverage samples that might contain trace amounts of naturally-occurring mycotoxins, the original sample mycotoxin concentrations should be corrected based on the pre-concentration and/or enrichment steps taken.

This collaborative study has demonstrated that analysis of five different mycotoxin types that would otherwise have involved 5 separate column clean-up protocols and mycotoxin separation methods, can successfully be consolidated into one LC/UV/fluorescence method with post-column derivatization. Perhaps, more important to developing countries is the advantage of a single column extraction for multiple toxins. Although results obtained in this study are only preliminary due to the insufficient number of trials and types of samples tested, they provide a strong indication of the potential this new approach possesses for multi-toxin analysis at high sample throughput.

Recently, R-Biopharm Rhône (Glasgow, Scotland) has developed prototype "(DONT2HT2Zea)" cartridge and Vicam (Watertown, MA) has developed an "ALL-IN-ONE" SPE column containing "(Aflatox inDONT2HT2FumonisinOchraZea)" antibodies. TTB collaborative mycotoxin projects with Waters and Varian have also successfully developed the LC/MS/MS analytical methods in conjunction with these new mixed-bed columns to analyze multi-toxin, including the two zearalenone metabolites (α- or β-zearalenol) in alcoholic beverages. It's inevitable that a growing number of new and different configurations of mixed-bed antibody-based and polymeric-based SPE column clean-up methods will be developed, including the rapid developments of the molecularly imprinted polymers (MIPs) column as worldwide regulations on various mycotoxins emerge.

SUMMARY

A collaborative project in multiple mycotoxins analysis has produced a technique capable of detecting trace-levels of five mycotoxin families in alcoholic beverages simultaneously. The new method relies on combining a multi-toxin mixed-bed immunoaffinity (IA) column, HPLC/UV/ fluorescence, post-column photochemical and *o*-Phthalaldehyde derivatizations, to analyze deoxynivalenol, aflatoxins B_1, B_2, G_1 and G_2, ochratoxin A, zearalenone and fumonisins B_1, B_2 and B_3. A prototype mixed-bed IA column, containing a 5-toxin (AflatoxinDONFumon isinOTAZea) monoclonal antibody resin support, used for multiple mycotoxins extraction, isolation and enrichment helped maximize throughput and minimize cost. The method detection limits were 20 µg/L for deoxynivalenol, 0.02 µg/L for aflatoxin G_2, 0.08 µg/L for aflatoxin G_1, 0.01 µg/L for aflatoxin B_2, 0.02 µg/L for aflatoxin B_1, 3 µg/L for fumonisin B_1, 3 µg/L for fumonisin B_2, 8 µg/L for fumonisin B_3, 0.14 µg/L for ochratoxin A, and 9 µg/L for zearalenone. Recoveries ranged from 71 to 112% for the spiked beer and rice wine samples.

ACKNOWLEDGEMENTS

Darsa Siantar wishes to dedicate this article to the late Dr. Sumer Dugar for his crucial role and support of this project. The authors also wish to thank Dr. Mary Trucksess (U.S. FDA, College Park, MD), Mark Benvenuti, Jim Krol and Joe Romano (Waters, Milford, MA), Nancy Zabe, Dr. Steve Powers (Vicam, Watertown, MA), Dr. Wentzel Gelderblom (PROMEC of Medical Research Council, Tygersberg, South Africa), Adiva Sotzsky, Norma Hill and Dr. Abdul Mabud (TTB, Beltsville, MD) for continuing to support this collaborative project.

REFERENCES

(1)   Abbas, H.K., Cartwright, R.B., Xie, W. and Shier, W.T. (2006) *Crop Prot.* **25**, 1

(2)   Zinedine, A., Brear, C., Elakhdari, S., Catano, C., Debegnach, F., Angelini, S., De Santis, B., Faid, M., Benlemlih, M., Minardi, V., and Miraglia, M. (2006) *Food Control* **17**, 868

(3)   Medina, A., Valle-Algarra, F.M., Mateo, R., Gimeno-Adelantado, J.V., Mateo, F. and Jimenez, M. (2006) *Int. J. Food Microbiol* **108**, 196

(4)   Simas, M.M.S., Botura, M.B., Correa, B., Sabino, M., Mallmann, C.A., Bittencourt, T.C.B.S.C., Batatinha, M.J.M. (2007) *Food Control* **18**, 404

(5)   Castellari, M., Fabbre, S., Fabiani, A., Amati, A., & Galassi, S. (2000) *J. Chromatogr. A*, **888**, 129

(6)   Otteneder, H. & Majerus, P. (2000) *Food Addit. Contam.*, **17**, 793

(7)   Pietri, A., Bertuzzi, T., Pallaroni, L., & Piva, G. (2001) *Food Addit. Contam.*, **18,** 647

(8)   Siantar, D.P., Halverson, C.A, Kirmiz, C., Peterson, G.F., Hill, N.R., & Dugar, S.M. (2003) *Am. J. Enol. Vitic.* **54,** 170

(9)   Visconti, A., Pascale, M., & Centonze, G. (1999) *J. Chromatogr. A*, **864,** 89

(10)   Visconti, A., Pascale, M., & Centonze, G. (2001) *J. AOAC Int.*, **84,** 1818

(11)   Zimmerli, B. & Dick, R. (1996) *Food Addit. Contam.*, **13,** 655

(12)   Scott, P.M. (1996) *J. AOAC Int.*, **79,** 875

(13)   Göbel, R. & Lusky, K. (2004) *J. AOAC Int.*, **87,** 411

(14)   Zabe, N., Wadleigh, E. and Cohen, B.A. (2006) in *Mycotoxins and Phycotoxins: Advances in Determination, Toxicology and Exposure Management*, Njapau, H., Trujillo, S., van Egmond, H.P. and Park, D.L. (Eds), Wageningen Academic Publishers, Wageningen, The Netherlands, 73

15. MYCOTOXIN REDUCTION AND DECONTAMINATION

Bruno Doko[1] and Douglas L. Park[2]

[1]International Atomic Energy Agency
Food and Environmental Protection Section
Joint FAO/IAEA Division of Nuclear Techniques in Food
and Agriculture
Wagramerstrasse 5, P.O. Box 100, A-1400
Vienna, Austria

[2]Center for Food Safety and Applied Nutrition
Food and Drug Administration
5100 Paint Branch Parkway
College Park, Maryland 20740

Mycotoxin contamination of agricultural commodities and the subsequent impact on consumers' (i.e., humans and animals) health as well as on national and international trade are increasingly recognized in both developed and developing countries. Developed countries enact regulatory levels to protect consumers from exposure to mycotoxins, but, in most developing countries, several agricultural commodities, including dietary staple foods, contain unacceptably high levels of mycotoxins. There are few regulatory guidelines and seldom are there economic incentives for high-quality products without mycotoxin contamination. If the food supply is limited, the mycotoxin hazard increases because more fungus-damaged, potentially mycotoxin-containing foodstuffs are consumed instead of being discarded, and malnutrition enhances a person's susceptibility to lower concentrations of food-borne mycotoxins (1). However, destroying contaminated products or diverting them to lower risk uses is not always practical (or affordable) and can seriously affect the food supply, particularly in economically deprived countries where, most of the time, nothing else is available for consumption.

Preventing mycotoxin contamination of agricultural commodities and/or, in a worst-case scenario, decontaminating or detoxifying contaminated food products promises to be the most practical approach in the effort to reduce mold and mycotoxin contamination. For consumer protection, any prevention strategy should focus on keeping mycotoxin levels in the food supply as low as possible. Therefore, preventive measures such as setting regulatory limits; monitoring for mycotoxin levels during harvesting, processing, and storage operations; and using decontamination procedures to reduce consumers' exposure remain the most appropriate approaches to the mycotoxin problem.

This chapter addresses (1) preventing mycotoxin production in agricultural commodities in the field (preharvest) and in storage (postharvest), and (2) postharvest decontamination procedures. The goal of decontamination (detoxification) is to degrade, destroy, and/or inactivate the mycotoxins or reduce amount of toxin per unit weight of a commodity to tolerable levels. Where possible, cost-effective detoxification technologies affordable to economically challenged regions of the world should be used or promoted.

PREVENTING MYCOTOXIN CONTAMINATION OF AGRICULTURAL COMMODITIES

Mycotoxin contamination of agricultural commodities occurs as a result of environmental conditions in the field as well as improper harvesting, storage, and processing operations. Failure to prevent fungal invasion and toxin formation (or production) in the field or in storage inevitably leads to an increased health risk from mycotoxins. Therefore, both pre- and postharvest strategies to prevent mycotoxin contamination are necessary to minimize consumers' exposure to harmful naturally occurring toxins in foods and feeds. The reader is referred to chapters

2 to 5 for more information on conditions leading to the formation of mycotoxins, and detailed discussions on health risks.

Preventing Mycotoxin Contamination in the Field

Sound mycotoxin management should begin in the field before harvest, where the toxigenic fungi first become associated with the crop and where the contamination process begins. Because there does not appear to be any single practical and effective technology available to prevent preharvest mycotoxin contamination, an integrated mycotoxin management approach seems appropriate (2, 3).

Good Agricultural Practices

Preharvest invasion of agricultural crops by fungi is governed primarily by plant host–fungus and other biological interactions (e.g., insects). Postharvest fungal growth is governed by the crop (nutrients) and by physical (temperature, moisture) and biotic factors (insects, etc). Field fungi such as *Fusarium* and *Alternaria* species require high relative humidity and water content and are not competitive in storage systems, which become dominated by storage fungi, particularly *Aspergillus* and *Penicillium* species that require minimal free water. Any preventive approaches to reduce mycotoxin contamination should be constructed upon the points stated above. Management technologies currently available for preventing preharvest contamination are irrigation and insect control. However, much more sophisticated strategies, such as biological control, the use of fungus- and toxin formation-resistant varieties, and expert computer systems that predict where and when contamination is most likely to occur need to be incorporated into the strategy.

Irrigation

Drought-stressed crops are plagued with a high incidence of mycotoxin contamination. Corn is subject to aflatoxin contamination during drought periods. Droughts and accompanying high temperatures during grain fill are conducive to fungal invasion and mycotoxin production in the kernel. Irrigation to decrease or reduce drought reduces the extent of mycotoxin production by fungi.

Insect Control

As reported by Cole and co-workers (2), insect damage is an important factor in predisposing crops to preharvest mycotoxin contamination. Therefore, effective insect control must be an integral part of preharvest mycotoxin management. Aflatoxin contamination is exacerbated by damage caused by the lesser corn stalk borer (*Elamopalpus lignosellus*) and termites in peanuts; pink boll worm (*Pectinophora gossypiella*) in cottonseed; navel orangeworm in almonds, walnuts, figs, and pistachios; and corn earworm (*Heliothis zea*) and European cornborer (*Ostrinia nubilalis*) in corn. Effective conventional insect control could have a major impact on reducing preharvest aflatoxin/mycotoxin contamination in several crops.

Preventing Mycotoxin Contamination During Storage

The development of mycotoxins in stored products can be avoided by preventing the growth of toxin-producing molds. Mold growth can be prevented by ensuring that moisture and temperature conditions favorable to growth and proliferation do not occur. Thus, as for the prevention of postharvest mold infection and subsequent mycotoxin contamination, the control of physical (temperature, moisture) and biotic factors (insects, interference competition) remains of great importance. In addition, in storage, the initial grain condition is critical as far as the fate of kernels is concerned. Good quality, clean, sound grain is easier to maintain in storage than physically damaged grain. Physically damaged kernels (i.e., broken kernels) are more prone to mold infection than whole kernels.

Moisture Control

Moisture content is by far the most important factor that affects the growth of microorganisms in stored grains. High temperatures, high relative humidity around the kernel, and kernel moisture between 14 and 30% (wet weight basis) are ideal conditions for fungal invasion of kernels.

Controlling the grain moisture or the water content of the environment can prevent the occurrence of mycotoxins such as the aflatoxins and fumonisins in stored grain. The growth of *Aspergillus flavus* and *Aspergillus parasiticus*, and subsequent aflatoxin production in storage, are favored by high humidity (>85%), high temperature (>25°C), and insect or rodent activity (4). *Fusarium moniliforme* has not been reported to grow in grain with moisture content below 18–20%, well above the recommended level of 13–14% for long-term storage. Therefore, no increase or production of the toxin should occur during storage if proper conditions of grain moisture and temperature are maintained (5). The formation (production) of zearalenone, an important mycotoxin in temperate and warm regions, takes place mainly before harvesting, but it may also occur postharvest if the crop is not handled and dried properly.

Postharvest handling of grains does present many more opportunities for controlling mycotoxin production. The main postharvest strategy involves drying commodities and keeping them dry (below 13% moisture). Artificial drying is commonly used today in developed countries. Although rural communities in Africa have no access to artificial drying facilities, appropriate drying and storage regimes are imperative. Sun drying is mostly used in developing countries. Drying to appropriate moisture levels to ensure safe storage in tropical climates is recommended to prevent fungal spoilage of the grain.

Insect Control

Insects as fungi vectors (carriers) throughout stored products are known to cause severe grain spoilage generally within long-term storage periods—i.e., ≥6 months. Left untreated, an insect infestation eventually leads to other storage problems. Insects give off moisture that can cause grain moisture contents to increase sufficiently to create a fungal problem. Fungal activity in turn raises temperature and results in an increased rate of insect reproduction, generating more moisture, and the cycle repeats at an ever-increasing rate. Broken grain is more vulnerable to insect attack, because some insects feed only on broken or cracked kernels. Recent investigations in four agroecological zones of Benin, West Africa (2), have demonstrated the influence of storage practices on aflatoxin contamination in corn. Factors associated with higher aflatoxin levels were storage time, insect damage, and storage structures. Therefore, factors that may help reduce the aflatoxin levels in stored corn include controlling insects by sorting out damaged ears, using appropriate storage insecticides, ensuring that farmers are aware of the risk that insects and aflatoxins present to their stored corn, and using proper storage containers.

DETOXIFICATION (DECONTAMINATION) OF AGRICULTURAL COMMODITIES

Food Processing

Practices for detoxifying (decontaminating) or inactivating mycotoxin-contaminated food and feeds comprise physical separation, thermal inactivation, irradiation, chemical treatments, the use of inorganic adsorptive compounds, and biological degradation.

Physical separation procedures are recognized as first approaches for reducing mold and mycotoxin contamination in agricultural commodities. These methods include cleaning (dry or wet cleaning), density segregation, handpicking, and manual and electronic sorting. Except for electronic sorting, these methods are popular in developing countries as a means for discarding bad looking ears/kernels, broken kernels, and insect- and fungus-damaged kernels. Cleaning procedures such as flotation and density segregation mechanisms are reported (7) to significantly reduce mycotoxin levels of products. The aflatoxin content of peanut kernels dropped from 300 to 20 µg/kg by using flotation as a separation mechanism. Similarly, a 25% reduction in the deoxynivalenol content of wheat flour was attained by cleaning and polishing the grain. In addition, cleaning by aspiration of spring wheat before milling reduces the level of the toxin in wheat flour. Furthermore, the patulin content of apples was reduced 93–99% by separating the rotten portions. Cleaning procedures that combine manual sorting and flotation of grain before milling are largely used in rural areas in Africa. The rejected portions are used mainly as poultry and ovine feed. The use of small-pore (mesh size) screens has been shown to considerably minimize the mycotoxin burden of grain. Screen cleaners that tumble and separate the fines, foreign materials, and physically damaged light-weight and infected kernels can be very effective in reducing the mycotoxin concentration in grain if used properly. The procedure can significantly decrease the toxin level in corn with a minimal 5–10% loss in grain weight or yield. Corn screenings mainly diverted to animal feed are known to contain the highest levels of fumonisins, causing severe documented outbreaks of animal diseases.

Industrial wet milling is a process widely used for cereals (corn, wheat, and sorghum) to produce several milling fractions (gluten, germ, fibers, and starch). The distribution of mycotoxins in these fractions varies. Wet milling produces starch free, or almost free, of mycotoxins. From fumonisin-contaminated corn, most detectable toxin ends up in the gluten, fiber, germ, and steep water fractions with very little or no fumonisins in the starch fraction (5, 8, 9). The major fractions produced by dry milling of corn are cornmeal, flour, and grits. Corn meal is the major staple food for most countries in Southern Africa. These products are also ingredients for breakfast cereals, snack foods, and baked goods such as cornbread and muffins. The bran fraction resulting from dry milling intended for animal feed has been shown to be highly contaminated with aflatoxins (10) and fumonisins (9).

Chemical and Physical Treatment

The detoxification strategy that has received the most attention is ammoniation of animal feedstuffs. Park and Price (12) showed, in a detailed review on reducing aflatoxin hazards with ammoniation, that the treatment was efficacious and the products and by-products were safe. No apparent toxicity in animals fed aflatoxin-contaminated ammonia-treated material was observed in the studies reviewed. The ammoniation process, using either ammonium hydroxide or gaseous ammonia, has been shown to reduce aflatoxin levels in corn, peanut meal cakes, and whole cottonseed products by more than 99% (12).

Primarily, two procedures are used today: a high-pressure and high-temperature (HP/HT) process used at treatment plants, and the atmospheric pressure and ambient-temperature (AP/AT) procedure that can be used on the farm. The HP/HT process involves treating the contaminated product with anhydrous ammonia and water in a contained vessel. The amount of ammonia (0.5–2%), moisture (12–16%), pressure (35–55 pounds per square inch), time (20—60 min), and temperature (80–120°C) varies with respect to the initial levels of aflatoxin in the product. The AP/AT process also uses anhydrous ammonia and water, which is sprayed on the product. The sprayed product is packed into a plastic silage-type bag, sealed, and held at ambient temperature (25–40°C) for 14–42 days or piled on a flat surface and covered with a tent for a similarly variable number of days. The holding time varies according to the ambient temperature; a lower ambient temperature requires a longer holding time. The amount of ammonia (1–5%) and moisture (16–26%) vary according to the initial levels of aflatoxin present (11). The ammoniation treatment products destined for animal feed. Several member countries of the European Economic Commission import ammonia-treated peanut meal on a regular basis. In the United States, the ammonia process is used in Arizona and California to reduce aflatoxin levels in cottonseed for dairy rations and to prevent the presence of aflatoxin M_1 in milk (11).

Heat treatment such as pasteurization has been shown to have minimal effect on aflatoxin M_1 content. This strongly suggests the need to prevent the presence of aflatoxin M_1 initially. Although fluctuations in aflatoxin M_1 concentration occur in the cheese production process and during preparation of cream and butter, balance studies revealed no losses of aflatoxin M_1 during the various processes (13).

The use of adsorbent materials such as activated carbon, zeolitic minerals, clay, etc. has been extensively studied (14). All have been reported to bind and/or remove mycotoxins. However, in recent studies for many of the adsorbents marketed today for binding mycotoxins, hydrated sodium calcium aluminosilicate clay a phyllosilicate clay commonly used as an anticaking agent in animal feeds, has been demonstrated to efficiently bind aflatoxin (15, 16) and is the most appropriate adsorbent for this class of toxins. Studies have also shown that hydrated sodium calcium aluminosilicate clay adsorbs aflatoxin B_1 with a high affinity and high capacity in aqueous solutions (including milk). By tightly binding the toxin, its bioavailability is markedly reduced, diminishing deleterious effects in animals (such as rats, chicks, turkey poults, lambs, and pigs) and decreasing the levels of aflatoxin M_1 in milk from lactating dairy cattle and goats. However, no positive effects were recorded with regard to the hyperestrogenic effects of zearalenone (15, 17).

In addition, biological control as a means to prevent mycotoxin production has generated notable success in approaches directed toward resistance to infection or reduction of mycotoxin contamination of grain (18). Genetic engineering is reported as a promising means of detoxifying mycotoxins (e.g., the fumonisins). This approach may provide innovative solutions to the problem of fumonisin in corn. Among the possibilities are genetically engineered resistance to *Fusarium* infection and genetic engineering approaches to detoxifying fumonisins in planta (5, 19). None of these approaches is affordable for developing countries.

SUMMARY AND RECOMMENDATIONS

The prevention and control of mycotoxin formation depend to a large extent on the commodity and on the fungus of concern, but some general principles apply. The approaches may be used before harvest, immediately after harvest, or during storage. For consumer protection, any prevention strategy should focus on keeping mycotoxin levels in the food supply as low as possible. Preventing mycotoxin contamination of agricultural commodities and/or, in a worst-case scenario, decontaminating or detoxifying contaminated food products are the best solutions for alleviating mold and mycotoxin contamination and ensuring consumer protection. Effective mycotoxin management requires an integrated approach that involves detection/diversion, separation/removal, sequestration, detoxification (decontamination), and prevention (2). Preventive measures such as setting regulatory limits for consumers (humans and animals), monitoring for mycotoxin levels at harvesting, and using processing and storage operations and decontamination procedures to reduce consumers' exposure to deleterious toxin levels remain the appropriate approaches to solve the mycotoxin problem. Table 16.1 presents effective and practical ways to prevent and/or control mycotoxin contamination of agricultural products. Moisture and insect control are the main approach to preventing fungal infection and subsequent mycotoxin formation in grain. As for food processing, operations such as physical removal/ separation of immature, damaged, or fungus-infected grain through screening and aqueous treatment (i.e., soaking/steeping or washing) and the milling process significantly reduce the mycotoxin levels in foods and alleviate to some extent the mycotoxin problem. However, thermal treatments (roasting, boiling, baking, frying, brewing) have a relatively smaller impact on mycotoxin destruction. Ammonia treatment and the use of hydrated sodium calcium aluminosilicate clay are the most effective detoxification means for mycotoxins.

Table 15.1. Suggestions for effective and practical procedures to prefent and/or control mycotoxin contamination of agricultural products

1. Control environmental factors that influence fungal infestation and/or growth
   - Moisture content of grain (12-13%)
   - Relative humidity (<70%)
   - Where feasible, choose varieties of grain that are resistant to insects, diseases, and mechanical damage[a]
   - Avoid drought and heat stress (irrigate where possible)
   - Avoid contact of grain with soil and other field debris
   - Apply phytosanitary treatment to eliminate insect damage

2. Control the physical condition/aspect of the grain
   - Minimize grain damage during harvest and crop handling
   - Harvest at maturity[b] and as soon as the moisture content allows minimum grain damage[c]
   - Dry the grain to 12–15% moisture content[d]
   - Thoroughly clean the grain to remove dirt, dust, and broken kernels and eliminate fungus-infected kernels, which are lighter than sound kernels

3. Control storage facilities
   - Thoroughly clean and keep storage facilities dry (e.g., warehouse, bins, sacks)
   - Store in insect- and rodent-tight structures

4. Physical removal procedures to control mycotoxin contamination of products
   - Floating separation, dry cleaning, and washing (to separate and eliminate fungus-infected kernels and screenings)
   - Sorting, handpicking

5. Decontamination and detoxification of products
   - Food processing
     - Washing and steeping
     - Milling process: wet and dry milling
     - Heating process: roasting, boiling, baking, frying, extrusion
   - Detoxification of products
     - Use of hydrated sodium calcium aluminosilicate in feed
     - The ammoniation process

[a]Because the conditions responsible for preharvest aflatoxin contamination of crops are characterized by high ambient/soil temperature and drought during the latter part of the growing season, prevention of aflatoxin contamination can be achieved by proper irrigation (2).
[b]Any damage to grain provides a route of entry for A. flavus and other toxin-forming fungi (20).
[c]Harvest when appropriate; crops left standing in the field are vulnerable to fungal infection.
[d]Do not exceed a 24- to 48-h period after harvest.

REFERENCES

(1)   Nelson, P.E., Desjardins, A.E., & Plattner, R.D (1993) *Annu. Rev. Phytopathol.* 31, 233

(2)   Cole, R.J., Dorner, J.W., & Blankenship, P.D. (1998) in *Mycotoxins and Phycotoxins—Developments in Chemistry, Toxicology and Food Safety.* M. Miarglia, H. van Egmond, C. Brera, & J. Gilbert (Eds), Alaken, Inc., Fort Collins, CO, 620

(3)   Lopez-Garcia, R., & Park, D.L. (1998) in *Mycotoxins in Agriculture and Food Safety.* K.K. Sinha & D. Bhatnagar (Eds), Marcel Dekker, New York, 407.

(4)   CAST (1989) *Mycotoxins: Economic and Health Risks,* Task Force Report 116, Council for Agricultural Science and Technology, Ames, IA

(5)   Munkvold, G.P., & Desjardins, A. (1997) *Plant Dis.* **81**, 556

(6)   Hell, K., Cardwell, K.F., Setamou, M., & Poehling, H.-M. (2000) *J. Stored Prod. Res.* **36,** 365

(7)   Park, D.L., & Liang, B. (1993) *Trends Food Sci. Tech.* **4**, 334

(8)   Riley, R.T., & Norred, W.P. (1999) *Preventing Mycotoxin Contamination.* FAO Food, Nutrition, and Agriculture Paper No. 2, Food and Agriculture Organization, Rome, Italy

(9)   Saunders, D.S., Meredith, F.I., & Voss, K.A. (2001) *Environ. Health Perspect.* **109**, 333

(10)   Njapau, H., Muzungaile, E.M., & Changa, R.C. (1997) *J. Sci. Food Agric.* **76**, 450

(11)   EC (1994) *Studies: Mycotoxins in Human Nutrition and Health,* EUR 16048 EN 1994, DG XII, European Commission, Brussels

(12)   Park, D.L., & Price, W.D. (2001) *Rev. Environ. Contam. Toxicol.* **171**, 139

(13)   Park, D.L., Lee, L.S., Price, R.L., & Pohland, A.E. (1988) *J. Assoc. Off. Anal. Chem.* **71**, 685

(14)   Phillips, T.D., Kubena, L.F., Harvey, R.B., Tasylor, D.R., & Heidelbaugh, N.D. (1998) *Poultry Sci.* **67**, 243

(15)   Grant, P.G., & Phillips, T.D. (1998) *J. Agric. Food Chem.* **46**, 599

(16)   Ledoux, D.R., Rottinghaus, G.E., & Bermudez, A.J. (2001) in *Mycotoxins and Phycotoxins in Perspective at the Turn of the Millennium.* W.J. de Koe, R.A. Samson, H.P. van Egmond, J. Gilbert, & A. Sabino (Eds), W.J. de Koe, Wageningen, 279

(17)   Lopez-Garcia, R., Park, D.L., & Phillips, T.D. (1999) in *Preventing Mycotoxin Contamination.* FAO Food, Nutrition, and Agriculture Paper No. 23, Food and Agriculture Organization, Rome, Italy

(18)   Bacon, C.W., Yates, I.E., Hinton, D.M., & Meredith, F. (2001) *Environ. Health Perspect.* **109**, 325

(19)   Duvick, J. (2001) *Environ. Health Perspect.* **109**, 337

(20)   Jacobsen, B.J., Bowen, Shelby, R.A., Diener, U.L., & Kemppainen, B.W. (1993) *Mycotoxins and Mycotoxicoses.* http: www.aces.edu/department/grai/ANR767.htm

16. MYCOTOXIN CONTROL: MONITORING PROGRAMS

Socrates Trujillo[1], Ezzeddine Boutrif[2], and Douglas L. Park[1]

[1]Center for Food Safety and Applied Nutrition
Food and Drug Administration
5100 Paint Branch Parkway
College Park, Maryland 20740

[2]Food and Agriculture Organization
Food Quality and Standards Service
C294 Via Delle Terme di Caracalla
0100 Rome, Italy

Monitoring is an exercise designed to continually check for the presence of and/or changes in the levels of a substance of interest. Mycotoxin monitoring programs are established as a risk management tool, but application of the information generated may not be restricted to risk management alone. Among the primary goals for establishing mycotoxin monitoring programs for foods and feeds are:

(a) To complement regulations (1, 2) where they exist and ensure that consumers of susceptible commodities are not exposed to levels that may have adverse effects on health,

(b) To acquire knowledge about the patterns of occurrence of the mycotoxins, and

(c) To detect and address contamination at early stages in the food-production chain (3–5).

Monitoring activities are an integral feature of food safety programs in several developed countries. On the contrary, for various reasons, monitoring activities are not a common feature in food safety initiatives in economically challenged nations. Food commodities produced in economically challenged nations are therefore of unknown quality and are usually subject to intense scrutiny when they enter international commerce (6–10).

FACTORS CONSIDERED WHEN ESTABLISHING A MONITORING PROGRAM

For monitoring to be successful, the nature of the target food or feed commodity, the agricultural production system and the type of processing, if any, must be considered. In addition, there should be a capability to identify and/or quantify the toxin(s) being monitored. To some extent, the desired level of accuracy of a quantification method is dictated by the intended use of the information acquired during monitoring. When only the presence or absence of a toxin needs to be known, simple, fast, and versatile analytical methods may be used. Several immunoassay techniques have been developed that are suitable for acquiring this type of information. Where near absolute accuracy is desired, a quantitative method requiring sophisticated equipment may be the choice. Thus, the laboratory should be sufficiently staffed and equipped to carry out such functions. Obviously, entire lots cannot be tested, so it is necessary to draw a portion of the material that represents the lot. As has been amply discussed in Chapter 6, establishing an appropriate sampling plan is an important component of any monitoring program.

A clear understanding of the production, storage, and marketing systems of the commodity to be monitored is vital. For instance, for a program targeted at monitoring aflatoxins in grains, the stage(s) at which the monitoring will be conducted should be clearly identified; i.e., will aflatoxin occurrence be monitored in the field, during shipment, during storage, or at all stages? The availability of resources, financial and otherwise, is always limiting (1,2,11,12). Therefore, monitoring programs should be established and executed in a cost-effective manner. Monitoring programs should not increase the cost of production to a level where the final product is unaffordable to consumers.

INDICATORS OF PROBABLE MYCOTOXIN CONTAMINATION

Where isolation and quantification of the actual toxin(s) are not feasible or the resources are inadequate, surrogate parameters may be monitored as indicators of probable mycotoxin contamination. It has been stated in Chapter 2 that fungi proliferate and produce mycotoxins under certain temperature and moisture conditions. Toxigenic fungal strains produce large quantities of mycotoxins during high moisture and temperature stress conditions (13–15). Similarly, grains harvested at high moisture are prone to fungal infestation and mycotoxin contamination during shipment. In addition, the moisture of kernels going into storage and the ambient temperature in the storage facility can provide an indication of the potential for mycotoxin occurrence. In some instances, the frequency, intensity, and duration of rainfall while susceptible commodities are in less-than-ideal storage conditions can be a warning sign for potential problems. Heavy rains at harvest or during storage lead to high moisture content in grains, which increases the possibility of vegetative fungal growth and mycotoxin production. Therefore, monitoring ambient temperature profiles and kernel moisture on the farm, particularly at critical stages of kernel development may provide clues to the likelihood of mycotoxin formation.

During harvest and transportation, grains may pick up fungal spores from combine harvesters and trucks. Reducing the spore load of equipment between shipments can considerably decrease the possibility of mycotoxin development during postharvest operations. Similarly, the presence of debris and filth around raw product storage areas, transportation facilities, and harvest equipment is an indicator of the potential occurrence of mycotoxins. Insect damage also increases at high ambient temperature and humidity levels. Hence, monitoring the cleanliness of farm implements, shipping containers, and storage facilities as well as the presence of insects and rodents may be one way to identify the potential for mycotoxin occurrence (16). Ultimately, the presence and quantity of the actual toxin(s) of interest should be ascertained.

ESTABLISHING A MONITORING PROGRAM ON THE BASIS OF THE HACCP CONCEPT

Incorporation of the hazard analysis critical control point (HACCP) concept for reducing risks associated with the presence of mycotoxins has been suggested (17–19). HACCP is a food safety approach that seeks to prevent food-borne hazardous events before they occur. The most probable high-risk stages (points) in a food production and manufacturing system are recognized, and preventive steps that can reduce, minimize, or eliminate potential risks are identified. Processors determine in advance how they will react if particular preventive steps fail and products or processes deviate from expected quality parameters. In the context of HACCP, monitoring ensures that the potentially high-risk stages in the commodity production process do not allow mycotoxin contamination to occur (18). Monitoring these control points is essential for detecting mycotoxin contamination (20).

MONITORING FOR MYCOTOXINS AT A CORN PROCESSING FACILITY

The scheme delineated below and illustrated in Figure 16.1 is hypothetical but can quite feasibly be transformed into reality. Incoming lots are tested and those in compliance with regulatory levels are placed in Storage A. Material contaminated within the range acceptable as animal feed is stored in Storage B. Lots whose contamination level is beyond that acceptable for human consumption or animal feed are placed in a third facility and shipped for industrial purposes, decontaminated, or destroyed.

Lots in Storages A and B are monitored for pest infestation (insects, birds, and/or rodents), broken kernels, diseased grain, moisture, and toxin content based on an appropriate monitoring program. The ambient temperature and humidity of the storage buildings are also monitored. During storage, toxin formation can occur if toxigenic fungi are present and allowed to grow in the commodity. It is important that further analyses, such as with immunochemical methods, be performed before the lots are ready for human consumption (21). Where occurrence or change in the level of toxin is suspected, confirmatory analysis with more advanced techniques such as thin-layer chromatography or liquid chromatography is performed. The end use of material in storage is determined by the mycotoxin content. In conceptualizing this scheme, the authors were aware that economically challenged nations may not have the luxury to segregate and divert contaminated commodities to nonhuman uses. However, highly contaminated lots can be reprocessed to reduce toxin levels.

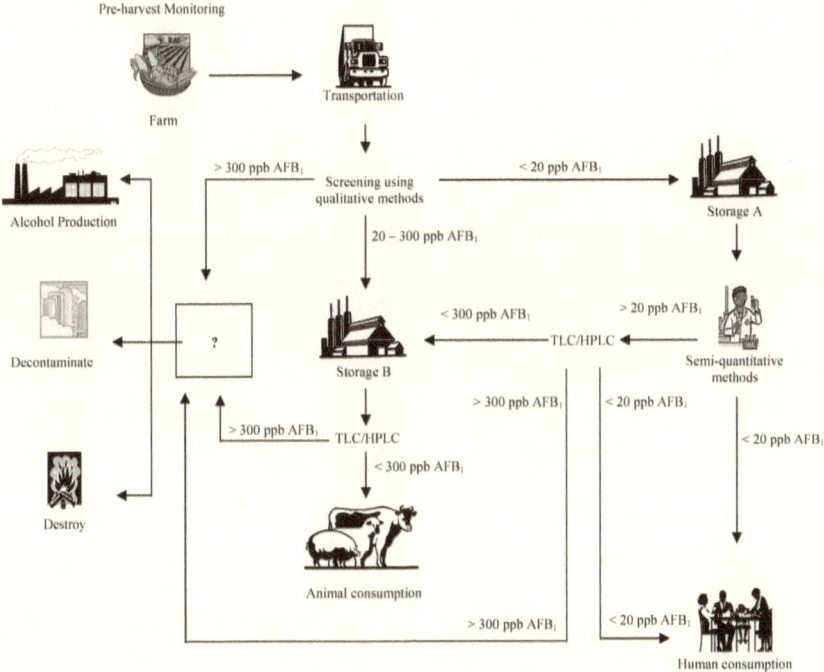

Figure 16.1. Proposed aflatoxin monitoring program for a large-scale corn processing operation.

SUMMARY

Mycotoxin-producing fungi may infect a product at any time during the production and manufacturing process. Monitoring for mycotoxins in susceptible commodities is necessary to limit the exposure of humans and animals, and to reduce the risks associated with consumption of contaminated commodities. Although it is ideal to monitor the presence of the toxin, factors that may trigger the production of mycotoxins need to be taken into account when monitoring programs are established. Monitoring programs should include inspection for insect damage, kernel moisture, and ambient temperature and humidity.

Mycotoxin detection procedures incorporated into a monitoring plan should include both rapid screening methods and quantitative confirmatory methods. Visual inspection is not effective as a rapid estimation tool for monitoring mycotoxin contamination. The incorporation of the concept of HACCP in a monitoring plan, in conjunction with other measures, would be helpful in reducing risks associated with the presence of contaminants in the final product.

REFERENCES

(1)   Hussein, H.S., & Brasel, J.M. (2001) *Toxicology* **167**, 101

(2)   IAEA.http://www.iaea.or.at/trc/myco-research.htm

(3)   Park, D.L., & Troxell, T. (2002) in *Mycotoxins and Food Safety*. M. W. Trucksess *et al.* (Eds), Kluwer Academic/Plenum Publishers, Gaithersburg, MD, 277

(4)   Pineiro, M., Dawson, R., & Costarrica, M.L. (1996) *Nat. Toxins* **4**, 242

(5)   Janardhana, G.R., Raveesha, K.A., & Shekar-Shetty, H. (1999) *Food Chem. Toxicol.* **37**, 863

(6)   Park, D.L. (1995) *Food Addit. Contam.* **12**, 361

(7)   Cardwell, K.F., Desjardins, A., Henry, S.H., Munkvold, G., & Robens, J. (2001) APSnet. http://www.apsnet.org/online/feature/mycotoxin/top.html

(8)   Dragacci, S., Grosso, F., Bire, R., Fremy, J.M., & Coulon, S. (1999) *Nat. Toxins* **7**, 167

(9)   Karki, T.B., & Sinha, B.P. (1998) in *Mycotoxin Prevention and Control in Foodgrains*. R.L. Semple, A.S. Frio, P.A. Hicks, & J.V. Lozare (Eds), Bangkok, Thailand. http://www.fao.org/docrep/x0036e/X0033E00.htm#Contents

(10)   Ontario Food Processing Research & Services Committee (2000) *Annual Report, January 2001*, Ontario, Canada

(11)   Zagrebenyev, D., Maier, D.E., & Woloshuk, C.P. (2001) Paper No. 01-6028. ASAE, St. Joseph, MI. http://abe.www.ecn.purdue.edu/~grainlab/research-rpts/current/dmitro-monitoring.html

(12)   Ioannou-Kakouri, E., Aletrari, M., Christou, E., Hadjioannou-Ralli, A., Koliou, A., & Akkelidou, D. (1999) *J. AOAC Int.* **82**, 883

(13)   Marks, B.P., & Stroshine, R.L. (1995) *J. Stored Prod. Res.* **31**, 343

(14)   Udoh, J.M., Cardwell, K.F., & Ikotun, T. (2000) *J. Stored Prod. Res.* **36,** 187

(15)   Hell, K., Cardwell, K.F., Setamou, M., & Poehling, H.M. (2000) *J. Stored Prod. Res.* **36,** 365

(16)   Doyle, M.E. (1997) *Food Research Institute Briefings*. University of Wisconsin–Madison. http://www.wisc.edu/fri/fusarium.htm

(17)   Park, D.L., Njapau, H., & Boutrif, E. (1999) *FNA* **23**, Food and Agriculture Organization (FAO), Rome, Italy, 49

(18)   European Mycotoxin Awareness Network. Fact Sheets on HACCP—Prevention and Control. http://www.lfra.co.uk/eman/fsheet3_1.htm

(19)   Herbst, K. (1994) *Food Sci. News. Hazleton Lab.* **53**, 1

(20)   FAO (2001) Joint FAO/WHO Food Standards Programme, 2001, CX/FC 01/22

(21)   Coker, R.D., Nagler, M.J., Defize, P.R., Derksen, G.B., Buchholz, H., Putzka, H.A., Hoogland, H.P., Roos, A.H., & Boenke, A. (2000) *J. AOAC Int.* **83**, 1252

17. MYCOTOXIN MONITORING AND ANALYTICAL SUPPORT: A PHILIPPINES PERSPECTIVE

Lyn A. Esteves

Bureau of Postharvest Research and Extension,
CLSU Compound,
Science City of Munoz, Nueva Ecija,
The Philippines 3120

MYCOTOXINS SITUATION IN THE PHILIPPINES

The aflatoxins are the only well studied mycotoxins in the Philippines. Their occurrence in agricultural commodities, such as corn, peanut and copra, as well as in products that are based on these commodities is well documented. With regard to other mycotoxins, there is dearth of information. There are studies suggesting possible occurrence of mycotoxins other than the aflatoxins because of the presence of mycotoxin-producing fungi belonging to the genus *Aspergillus*, *Fusarium* and *Penicillium*. However, no extensive surveys have been conducted to determine the nature and extent of contamination by mycotoxins produced by these fungi.

INFORMATION CAMPAIGN ON MYCOTOXIN PREVENTION AND CONTROL

Local and international research on aflatoxin has generated voluminous information on various aspects of the problem. To disseminate available information, the Bureau of Postharvest Research and Extension (BPRE) has launched an Information Campaign on Mycotoxin Prevention and Control. The effort is spearheaded by the Applied Communication Division of BPRE. The objective of the campaign is to increase awareness of stakeholders in the corn industry on the various aspects of aflatoxin contamination. The campaign is focused on corn because, among susceptible commodities, corn presents the highest risk from aflatoxin contamination. It is the staple food of 20 percent of the Philippine population and widely used as animal feed. The campaign is envisioned to lead to a corn industry that understands the aflatoxin problem and positioned to reduce contamination through application of recommended pre- and post-harvest practices and adoption of appropriate technologies.

The information campaign has two components, namely: communication support, and training. The communication support component involves the production and dissemination of communication support materials through various channels such as print, broadcast and audio-visual media. Examples of the communication support materials and their corresponding target clients and contents are illustrated in Table 17.1. An innovative School-on-the-air (SOA) program will also be launched. The SOA program is a strategy to educate farmers through radio. It is comprised of 8 sessions and scheduled to be aired from 5:30-6:00 AM every Tuesdays at DWPE, a local radio station in Isabela, which is one of the major corn-producing areas in the country.

Table 17.1. Communication support format, content and target audience

Communication support	Clients	Contents
Flyer, brochure, TV plugs, wall newspaper, calendar, exhibits	General public	General information on mycotoxins
Primer, pamphlet, techno flyers, "How-to"	Farmers, millers, traders	Occurrence of mycotoxins in food and feeds, health risks associated with mycotoxins, prevention and control of aflatoxins in corn
School-On-The-Air (SOA)	Farmers	Pre- and post-harvest practices and technologies that can prevent or control aflatoxin contamination

The training component involves training of regional information officers and regional corn coordinators. These groups of government agricultural officers can serve as trainers for future mycotoxins training courses. In addition, the regional corn coordinators are responsible for planning and implementing the Corn Program of the Department of Agriculture, which is focused not only on increasing corn production but also on production of quality and safe corn. Knowledge of the various aspects of aflatoxin contamination in corn enables them prepare and implement plans that will help address the aflatoxin problem in their respective regions.

The training of regional information officers and other experts in information dissemination is scheduled for June 8-10, 2004 while that of the regional corn coordinators is scheduled for July 27-30 of the same year. A training module was prepared following the general framework of the International Workshop on Mycotoxins held on July 22-26, 2002 at the U.S. Food and Drug Administration in College Park, Maryland, USA. The training course for information officers is divided into five modules:

Module 1:   General Overview on the Information Campaign Program and Corn Postproduction Industry

Module 2:   Overview of Mycotoxin Problems in Corn

Module 3:   Surveillance and Analysis of Mycotoxins

Module 4:   Prevention and Control of Aflatoxin

Module 5:   Hands-on exercises/workshop

The specific topics covered in each module are given in Table 17.2.

Table 17.2. Content of a Regional Information Officers' training module.

Module	Topics included
Module 1: General overview on the information campaign program and corn post production industry	• Mycotoxins information campaign program • Mycotoxins briefing materials to be developed • Briefing on action planning • Overview of corn post production industry
Module 2: Overviw of the mycotoxin problem in corn.	• Definition of terms • Conditions leading to occurrence of mycotoxins • Fungi and the toxins they produce • Human health implications • Animal health implications • Socio-economic implications
Module 3: Analysis and surveillance of mycotoxins	• Methods of analysis and sampling • Requirements for the establishment and management of mycotoxins laboratory
Module4: Prevention and control of aflatoxin	• Corn aflatoxin control system (MACS) • Regulatory control • Available technologies to reduce contamination
Module 5: Hands-on and workshop	• Field trip to feed mill; collection of samples • Lab exercises on sample preparation; aflatoxin testing using rapid test kits • Action planning on preparation of information materials • Presentation of action plans

The training module for the regional corn coordinators is basically similar but their action plan is on the conduct of training activities in their respective regions. An interdisciplinary pool of subject matter specialists from BPRE and other agencies involved in mycotoxins research shall serve as resource speakers in the training courses.

Future activities include the production of standardized lecture materials for training of trainers to harmonize mycotoxins training in the Philippines. The handbook of the International Workshop on Mycotoxins will be used as guide in the preparation of said materials.

MYCOTOXIN MONITORING ACTIVITIES IN THE PHILIPPINES

Monitoring by Regulatory Agencies

Human Food

The knowledge that mycotoxins can have serious effects on human and animal health has prompted many countries to establish mycotoxin regulations in food and feeds (1). However, owing to the scarcity of information on other mycotoxins, aflatoxin is the only mycotoxin regulated in the Philippines. Aflatoxin is of major concern in corn, peanuts, and copra. Corn is mainly used as animal feed; it is also the staple food of about 20 percent of the Philippine population and a major raw material for snack foods. Peanuts are mainly consumed as snacks while copra is primarily used as feedstuff and is a major export commodity. Although there is a clear agreement on the need for aflatoxin regulatory control measures in both food and feeds, aflatoxin regulation is applied only to human food.

Regulation of mycotoxins in foods in the Philippines is based on the general provisions of the Food, Drug and Cosmetic Act (Republic Act 3720, as amended by Executive Order No. 175) and The Consumer Act of the Philippines (Republic Act 7394). These Acts prohibit the manufacture, sale, offering for sale or transfer of any adulterated food, which is defined as "food bearing or containing any poisonous or deleterious substance which may render it injurious to health." Based on this definition, any food containing mycotoxins in quantities that render it injurious to health can be considered adulterated. The Bureau of Food and Drugs (BFAD) in the Department of Health is the agency mandated to administer and enforce the laws pertaining to these acts.

There are no locally developed national standards on mycotoxins in the country. Instead, the BFAD follows recommendations of the Codex Commission. Hence, the total aflatoxin limit for human foods in the Philippines is set at 20 µg/kg. To determine compliance with this standard, BFAD regularly monitors manufactured corn- and peanut-based food products. Samples are collected by Food and Drug Regulatory Officers during routine inspection of food processing plants and also from retail stores. Aflatoxin analysis is performed by the Laboratory Services Division (LSD) of BFAD. The LSD uses an enzyme linked immunosorbent assay (ELISA) method with an ELISA reader. It applies analytical quality assurance principles, including participation in proficiency testing.

Animal Feeds

The relevant legislation for regulating mycotoxins in feeds is the Livestock and Poultry Feeds Act (Republic Act 1556, as amended by Presidential Decree No. 7). The rules and regulations pertaining to this act are administered by the Department of Agriculture through the Bureau of Animal Industry (BAI). The BAI has a mycotoxins laboratory that also analyzes samples brought in by feed millers and livestock and poultry farmers incapable of conducting their own mycotoxin analysis. The BAI laboratory uses the Natural Resources Institute of the United Kingdom (NRI) Bond Elut method with high performance thin layer chromatography (HPTLC) for aflatoxin and the ELISA method with an ELISA reader for aflatoxin and other mycotoxins. The BAI cannot, however, sustain regular monitoring of aflatoxin in feeds mainly due to the high cost of solvents and reagents.

Monitoring by the Food and Feed Industry

Despite the non-implementation of aflatoxin regulations in animal feeds, almost all large feed manufacturers, who account for more than 50% of local feed production, routinely monitor aflatoxin in raw materials and/ or finished products. The industry standard is 50 μg/kg total aflatoxin for raw materials such as corn and copra. The 50 μg/kg level is based on the Philippine Society of Animal Nutritionists (PHILSAN) guidelines. The standard for mixed feeds and any other finished products is 20 μg/kg total aflatoxin. Some large-scale feed manufacturers also monitor imported raw materials such as soy beans and wheat for T-2 toxin, zearalenone, and deoxynivalenol. The "self regulation" of the feed industry may be due in part to the recognized deleterious effects of high mycotoxin levels in poultry and livestock, which in some cases have resulted in litigation between the feed manufacturers and the poultry and livestock industry.

Generally, only large food manufacturers have in-house facilities for aflatoxin testing and routinely monitor raw materials and/or finished products for aflatoxin. The current practice of other food manufacturers is to sample only lots that are suspected of being contaminated with aflatoxin, usually based on physical appearance. Samples of the suspected feed are sent to BFAD-recognized laboratories for aflatoxin analysis.

Based on an informal survey, it was found that most food and feeds laboratories use ELISA methods with an ELISA reader for mycotoxin analysis. A few laboratories use immunoaffinity column methods with a flourometer or a comparator card. Sampling procedures vary, with lot samples ranging from as little as 250 grams to as much as 5 kilograms. Samples are collected by taking increments from exposed bags in a lot, which is usually a truckload, or by systematic random sampling using spears.

Advances and Constraints in Mycotoxin Monitoring

Effective monitoring requires reliable methods of sampling and analysis. Sampling is particularly crucial because of the non-homogenous distribution of mycotoxins in contaminated lots. To ensure that the test results accurately reflect the mycotoxin concentration of the contaminated lot, samples must be representative of the lot and of sufficient size to compensate for the uneven distribution of mycotoxins. Hence a harmonized sampling scheme that is based on sound statistical principles is highly desirable. However, practical considerations must also be taken into account because there is little benefit in adopting a procedure that is too laborious and requiring a large amount of sample that presents handling and preparation problems. The 1993 FAO Technical Consultation on Sampling Plans for Aflatoxin Analysis in Peanuts and Corn evaluated a range of sampling schemes for countries to review and choose from. Sampling schemes for peanuts and corn based on statistical principles as well as practical considerations were evaluated using operating characteristics (OC) curves and lot distributions (2). These OC curves and lot distribution attributes describe the goodness of a sampling plan for inspecting a single lot with a given concentration. From the OC curve and the lot distribution, the total number of lots accepted and rejected and the proportion of correct and incorrect decisions can be computed for a given sampling plan.

The OC curves were developed using the negative binomial distribution, which is characterized by high probability of kernels with zero aflatoxin and low probability of kernels with high aflatoxin levels. To determine the suitability of the negative binomial distribution in describing aflatoxin distribution under Philippine conditions, a study on the distribution of aflatoxin in farmers' shelled corn lots was conducted (unpublished report). This study showed that the negative binomial distribution adequately fitted the aflatoxin distribution in the lots that were tested. Hence, the OC curves developed by the FAO Technical Consultation, coupled with lot distributions using Philippine data, became the basis for selecting a sampling scheme.

The high cost of analysis continues to be a major deterrent to routine monitoring for mycotoxins in the Philippines. To address this concern, Arim and coworkers at the Food and Nutrition Research Institute developed an aflatoxin field test kit for copra and copra meal (3). The method is based on adsorbent technology and utilizes an indigenous adsorbent material from volcanic ash. This test kit was adapted for use in corn (4) and, together with two commercially available aflatoxin test kits, was subjected to interlaboratory testing within the Philippines involving 12 government laboratories (5). Furthermore, Espino and coworkers at the University of the Philippines Los Baños-National Institute of Molecular Biology and Biotechnology (UPLB-BIOTECH) similarly developed a diagnostic kit for aflatoxin B_1 using monoclonal antibodies (6).

FUTURE STRATEGIES

Clearly, there is still a need to develop simple, rapid, selective, and inexpensive methods of mycotoxin analysis. One emerging analytical technique is the use of molecularly imprinted polymers (MIPs) as substitutes for antibodies in assays or sensors. Like antibodies, MIPs are highly selective. The advantages of MIPs over antibodies are inherent stability, low cost, and ease of preparation. MIPs for deoxynivalenol, zearalenone, and ochratoxin have been developed and evaluated in solid phase extraction (SPE) and sensor formats (7). Another promising technology is the use of near-infrared spectroscopy in detecting mycotoxins. Pearson et al., (8) developed a grain sorter to detect aflatoxin and fumonisin in corn kernels using two bands of infrared light. The commercialization of the techniques should be pursued with considerable urgency.

Likewise, studies on the distribution of aflatoxin and other important mycotoxins in commodities other than those that have been investigated should be pursued so that appropriate sampling plans could be developed. Aside from aflatoxin, suspected fumonisins contamination in corn and ochratoxins contamination in coffee, are gaining interest in the country.

Although many of the aflatoxin test kits available on the Philippine market have undergone validation that demonstrated their reliability, laboratories need to verify the performance of these kits prior to use. In addition, they need to apply other quality assurance procedures including the use of certified reference materials and regular participation in proficiency testing schemes. However, certified reference materials and regular participation in externally organized proficiency testing entail considerable cost and may not be practical for economically-challenged countries like the Philippines. A pragmatic approach is for a national agency to lead an analytical quality assurance program that would include: the provision of reference materials to various mycotoxins laboratories, conduct of regular proficiency testing and, training courses on methods of analysis and sampling including analytical quality assurance. A number of government laboratories in the Philippines have the capacity and competence to lead such a program. What is lacking in many of these laboratories is a formal accreditation based on international standards. In the absence of the accreditation, establishing links and cooperation with competent laboratories abroad should increase the level of confidence in data generated by these laboratories. International organizations like the FAO could foster such collaborations.

SUMMARY

Although cognizant of the risks posed by other mycotoxins, the Philippines has limited resources to address all mycotoxins concerns. Minimizing the risks associated with mycotoxin contamination requires more than just monitoring. However, if monitoring is successful, consumption of food and feeds that are contaminated with mycotoxins would be substantially minimized. Hence, the challenge to the analytical community is to develop simple, fast, inexpensive but reliable methods for sampling and analysis, that will ensure effectiveness and sustainability of mycotoxin monitoring programs especially in economically challenged countries like the Philippines. It is also a challenge to the analytical community in developing countries to espouse and practice quality assurance procedures to ensure that the data they generate is of high quality and can be the basis of sound food safety decisions.

ACKNOWLEDGMENTS

Gratitude is extended to BFAD and BAI, and the food and feed manufacturers in the Philippines who shared information regarding their mycotoxin monitoring activities, and Ms. Miriam Acda, Acting Director II of BPRE, for valuable comments and suggestions on this manuscript.

REFERENCES

(1)   FAO (2004) *Worldwide Regulations for Mycotoxins in Food and Feed*, Food and Agriculture Organization, Rome, Italy

(2)   FAO (1993) *Sampling Plans for Aflatoxin Analysis in Peanuts and Corn*. Food and Nutrition Paper 55., Food and Agriculture Organization , Rome, Italy

(3)   Arim, R., A. Aguinaldo, T. Tanaka, and T. Yoshizawa. (1999a) *J. AOAC Int.* **82**, 877.

(4)   Arim, R., A. Aguinaldo, T. Tanaka, and T. Yoshizawa. (1999b) *Mycotoxins*, **48**, 53

(5)   Esteves, L., V. Lumba, R. Arim, and C. Ferolin. (2000) *Validation of Potential Rapid Methods for Aflatoxin Detection in Corn*. Terminal Report. BPRE, Muñoz, Nueva Ecija, The Philippines.

 (6)   Fernandez, R. (2003). *Bohol Now Producing Export Quality Peanut Kisses*. Philippine Star, Nov 22, Manila, The Philippine

(7)   Molinelli, A., R. Weiss, J. Mahony, K. Nolan, M.R. Smyth, R. Krska and B. Mizaikoff. (2004). *Molecularly Imprinted Polymers for Mycotoxins: Characterization and Optimization of Synthetic Receptors for Improved Beverage Analysis*. Poster presented at the XI IUPAC Symposium on Mycotoxins and Phycotoxins, Maryland, USA

(8)   Pearson, T.C., Wicklow, D.T., Pasikatan, M.C. (2004) *Cereal Chem.* **81**,:490

18. LOWER FUMONISIN LEVELS IN THE GRAIN OF *Bt* CORN

Bruce Hammond[*], Dominque Melcion[1], Pierre-Yves Kergoat[2], Juan J. Sequeira[3], Jacqueline Cea[4], Hamit Esin[5] Fahri Tatli[6] and Gianfranco Piva[7]

*Monsanto Company,
Building O3F,
800 North Lindbergh Blvd,
St. Louis, MO. 63167, USA

[1]INRA, UR 783 UPCM,
Physico-Chimie des Macromolecule
INRA rue de la Geraudiere,
BP71 627, 44316, Nantes, France

[2]Monsanto Agriculture France SAS
Hybritech/Asgrow,
1 Rue Jacques Monod, Europarc du Chene
Bron Cedex, 69673, France.

[3]Monsanto Argentina SAIC
Maipu 1210- 10th Floor,
Buenos Aires C1006 ACT, Argentina

[4]Mycotoxin Department,
Technological Laboratory of Uruguay,
Ave Italia, 6201, CP 11500, Montevideo
Uruguay,

[5]Monsanto Gide ve Tarim Tic.
Fahrettin Kerim Gokay Cad. No:22
A Blok Kat: 2 Altunizade-Istanbul/Turkey

[6]Crop Protection Institute of Ministry of Agriculture and
Rural Affairs
Koprukoy-Adana, Turkey

[7]Istituto di Scienze degli Alimenti e della Nutrizione,
Universita Cattolica del Sacro Cuore
Via Emilia Parmense, 84 29100 Piacenza, Italy

Corn (*Zea mays* L.) can be infected with fungi that produce toxic secondary metabolites called mycotoxins. *Fusarium verticillioides* (Sacc.) Nirenberg (synonym = *F. moniliforme* J. Sheld.) and *F. proliferatum* (T. Matsushima) Nirenberg which produce fumonisin mycotoxins are the most common fungi that infect corn wherever it is grown. Various environmental factors such as insect damage, heat and drought stress, and genetic susceptibility predispose corn plants to infection with fungi (1). Dietary exposure to fumonisins can cause a variety of adverse health effects in farm and laboratory animals (2). Epidemiological studies suggest a link between high dietary intake of fumonisins and elevated rates of liver and/or esophageal cancer in certain regions of Africa, China, Italy and Brazil (2, 3). As a consequence, regulatory agencies have recommended limits for fumonisin contamination of corn grain intended for animal feed and human food use (2).

Control of insect pest damage of corn can reduce fungal infection since insect feeding provides ports of entry for fungi. Some insect pests also serve as vectors for fungal infection (4). An effective insect pest control strategy has been developed with the introduction of coding sequences for the Cry1Ab protein derived from *Bacillus thuringiensis (Bt)* into corn plants (event MON 810, YIELDGARD CORNBORER, trademark of Monsanto Company) (5). The Cry1Ab protein controls lepidopteran insect pests such as the European Corn Borer (ECB), *Ostrinia nubilalis*

Hübner, the most important stalk-boring and ear damaging insect pest of corn in the United States Corn Belt (6) The CaMV 35S gene promoter enables constitutive expression of the Cry1Ab protein throughout the growing season, thus providing season-long protection against corn borers. The *B. thuringiensis*-based microbial pesticides that contain Cry proteins such as Cry1Ab have been used commercially in agriculture for over 40 years to control larval insect pests (5, 7). They are safe because their insecticidal mode of action is highly specific against target lepidopteran insect pests. The Cry 1Ab protein has no activity against non-target organisms such as mammals and birds (5,7).

Munkvold and co-workers were the first to report that corn hybrids protected with the Cry1Ab protein had significantly lower fumonisin mycotoxin levels in the grain (8). This was most evident in corn plants that expressed Cry1Ab protein constitutively during the growing season. Additional field trials have been conducted in countries that permitted field testing of *Bt* hybrids to assess their impact on fumonisin levels under local conditions. Results are presented from field trials in Italy, France, Turkey and Argentina that compared the levels of fumonisins in *Bt* hybrids derived from the MON 810 event with their near-isogenic controls.

MATERIALS AND METHODS

Turkey

Field trials with *Bt* corn (derived from event MON 810) were carried out in Adana Province, in the East Mediterranean Cukurova region where corn is planted as a second crop after wheat. Damage from corn boring insect pests is most severe in the second corn crop. The hybrids used for the study were DK626 *Bt*, its near-isogenic control (DK626) and a conventional hybrid traditionally grown in the region. The trials were carried out under conditions of natural insect infestation. Trials in 2000 and 2001 were set up using a randomized split plot design with four replicates. Each of the four blocks was divided into two sub-blocks. Four sub-blocks received three applications of the insecticide lamba-cyhalotrine at two to three week intervals, starting the third week of July. The other sub-blocks were not treated with an insecticide. Test plots consisted of eight corn rows 20 m in length. At harvest, mycotoxin concentrations were compared across treatments. ELISA test kits (quantitative kit for fumonisin, Veratox, Neogen Corp., Lansing, MI) were used to determine fumonisin levels in accordance with the kit instructions.

Italy

In 1999, field trials were carried out in northern Italy at 30 different locations with four Bt hybrids of different genotypes (derived from event MON 810) and their respective near-isogenic controls. The trials were carried out under conditions of natural insect infestation as there is significant ECB infestation in these regions. Approximately 93 samples of maize grain were randomly collected from one to 10 weeks post harvest from the various locations and sent to the School of Agriculture, U.C.S.C. Piacenza, Italy, for analysis using published procedures (9).

France

During 1997 to 1999, the levels of fumonisin mycotoxins were measured in *Bt* hybrids (derived from MON 810) and their near-isogenic controls at 25 field trial locations in France. The majority of the sites were in the southwest of France where ECB is normally present. The trials were carried out under conditions of natural insect infestation. In non-replicated field trials, the plot size was a minimum of 20 m x 8 rows. In replicated field trials, there were four randomized blocks in two plots (*Bt* and non *Bt*), in plots 12 meters by six rows. For non-replicated trials, grain was collected from 20 consecutive plants in two adjacent rows at four locations in the plot. For replicated trials, grain was collected from 20 consecutive plants in two adjacent rows for each replicate. Grain harvested in 1999 trials was analyzed for fumonisins at INRA, Nantes France according to published methods (10). Grain samples harvested in 1998 were analyzed for fumonisins at IEEB (European

Institute of Environment at Bordeaux) using established methods (9). For grain harvested in 1997, samples were analyzed at AGPM (French Corn Growers Association) using ELISA methods (Diffchamb France, Transia® plate Fumonisins).

Argentina

A *Bt* hybrid (DK696) derived from event MON 810 and its near-isogenic control (DK696) were grown in 57 different locations in Buenos Aires Province in Argentina during 2001. This province is an important corn growing region of Argentina and corn pests such as *Diatraea saccharalis* (corn borer) and *Helicoverpa zea* (ear worm) are frequently found in this region. The field trials were carried out under conditions of natural insect infestation. Grain samples were collected from each of the sites at harvest and submitted for fumonisin analysis at the Laboratory Technology of Uruguay (LATU) in Montevideo using published methods (11).

RESULTS AND DISCUSSION

As shown in Table 18.1 and Figures 18.1-18.3, many of the field locations had lower fumonisin levels in *Bt* hybrids compared to their near isogenic controls. These results confirm that insect feeding is an important contributor to fumonisin contamination of grain under conditions where there is significant corn borer presence. Protecting grain against insect feeding damage can reduce the opportunity for fungi to infect kernels. *Bt* hybrids have been reported to have lower fungal contamination based on measurements of ergosterol levels in grain (10, 12).

When the levels of fumonisins were averaged across all sites for trials in each country (Table 18.2), the *Bt* hybrids had reductions ranging from 61 to 97% compared to their near isogenic controls. In field trials carried out across the United States in 2000 and 2001, fumonisins levels were approximately 50% lower when averaged across all locations (180 total sites) for each year (13).

Table 18.1. Lower fumonisin levels in *Bt* hybrid DK 626 grown in Turkey 2001 and 2002

Varieties	Fumonisin Level (mg/kg)				Average fumonisin level (mg/kg)
	Field trial 2001				
	1st Rep.[c]	2nd Rep.	3rd Rep.	4th Rep.	
DK626 Bt (-)[a]	3.5	2.0	2.5	2.0	2.5
DK626 Bt (+)[b]	3.7	2.3	1.8	2.8	2.6
DK626 (-)	16.8	15.1	18.3	19.9	17.5
DK626 (+)	14.1	15.6	15.4	17.4	15.6
Conventional (-)	18.8	16.7	18.0	18.8	18.1
Conventional (+)	17.2	16.4	15.0	17.9	16.6
	Field Trial 2002				
DK626 Bt (-)	0.8	1.0	0.4	0.8	0.78
DK626 Bt (+)	0.5	0.4	0.8	0.7	0.63
DK626 (-)	17.3	18.7	18.2	12.8	16.75
DK626 (+)	5.5	10.6	18.0	16.7	12.7
Conventional (-)	18.7	17.5	16.6	18.0	17.7
Conventional (+)	12.8	12.2	15.6	17.8	14.6

[a]Untreated pots(-)
[b]Insecticide treated plots (+)
[c]Cobs were wounded artifically.

Table 18.2. Percent reduction in fumonisin levels observed in field trial studies by country

Country	Year	Sites	Non *Bt* hybrids mean fumonisin (mg/kg)	*Bt* hybrids mean fumonisin (mg/kg)	% Reduction in mean fumonisin
Turkey	2001	2	17.5	2.6	85
France	1997-1999	26	1.0	0.03	97
Italy	1999	30	2.9	0.35	88
Argentina	2001	57	5.03	1.95	61

Significant fumonisin contamination of corn intended for human consumption continues to occur as evidenced by recent reports of its presence in corn meal (mean level 2,200 µg/kg) and corn based infant food (14). Other scientists are investigating possible links between fumonisin exposure in early pregnancy and increased incidence of neural tube defects in newborns (15). Other dietary factors such as folic acid deficiencies (16) have also been linked to neural tube defects. Fumonisin and aflatoxin co-contamination of corn grain has also been reported in some locations (17), which may be of concern since fumonisins promote the liver carcinogenic activity of aflatoxin in animal models (18, 19). The potential impact of fumonisin exposure on human health is being actively investigated.

CONCLUSION

Fumonisin contamination occurs wherever corn is grown. Field trial experiments in locations where corn borers are prevalent have consistently found lower fumonisin contamination in the grain of *Bt* hybrids. At some locations, the reduction in fumonisin levels was large enough to make the difference between an unacceptable crop and one that was safe for consumption. The cultivation of *Bt* hybrids can be a useful tool in the reduction of potentially harmful mycotoxins in corn grain, but is not a complete solution. Based on the findings reported here, the next generation of *Bt* hybrids that control a broader spectrum of insect pests, may be even more effective in lowering mycotoxin contamination of corn grain.

SUMMARY

Insect feeding injures corn kernels creating ports of entry for fungi that produce ear rot and mycotoxins. Protection of corn kernels against insect feeding could reduce fungal infection and mycotoxin contamination. This has been demonstrated with a new variety of corn hybrid (event MON 810) genetically modified to protect corn against the European corn borer, a serious pest of field corn. The coding sequence for Cry1Ab protein derived from *Bacillus thuringiensis* (*Bt*) was introduced into the corn genome. Cry proteins are the active ingredients of *Bt* microbial insecticides that have been safely used on agricultural crops around the world for forty years. The Cry1Ab protein is produced throughout corn tissues providing season-long protection against corn borers.

Scientists who first evaluated *Bt* corn in field tests in Iowa found less damage from corn borers and lower ear rot and fumonisin mycotoxin levels in the grain. These initial observations were followed by more extensive field trials conducted across the United States and in other countries (France, Italy, Argentina and Turkey) at locations with significant corn borer pressure. Fumonisin levels in *Bt* hybrids were compared to near isogenic controls. When averaged across all sites for two consecutive years, fumonisin levels for several different Bt hybrids were approximately 50% lower in the United States (180 sites). In Argentina, results during one season with one *Bt* hybrid tested at over 50 locations found fumonisin levels on average 60% lower. In Turkey, one *Bt* hybrid tested for two consecutive years had fumonisin levels 85% lower than controls. Field trials in France (four + years) and Italy (one year) at multiple locations also found consistently lower fumonisin levels in *Bt* hybrids. At some locations, the reduction in fumonisin levels was large enough to make the difference between an unacceptable crop and one that was safe for consumption.

REFERENCES

(1)   Miller, J. D. (2001) *Environ. Health Perspect.* **109** (suppl. 2), 321

(2)   CAST, (2003). *Mycotoxins, Risks in Plant, Animal, and Human Systems,* Task Force Report No. 139, Council for Agricultural Science and Technology, Ames, Iowa, 54.

(3)   van der Westhuizen, L., Shephard, G.S., Scussel V.M., Costa L.L., Vismer, H.F., Rheeder, J.P., & Marasas, W.F., (2003) *J. Agric. Food Chem.* **51,** 5574

(4)   Sobek, E.A. & Munkvold, G.P., (1999) *J. Econ. Entomol.* **92,** 503

(5)   Betz, F. S., Hammond, B.G., & Fuchs, R. L ., (2000) *Regul. Toxicol. Pharmacol.* **32,** 156

(6)   Pilcher, C.D., Rice, M.E., Obrycki, J.J., & Lewis, L.C. (1997) *J. Econ. Entomol.* **90**, 669

(7)   IPCS., 1999. *Environmental Health Criteria 217: Bacillus thuringiensis,* International Programme on Chemical Safety, World Health Organization. Geneva, Switzerland 36

(8)   Munkvold, G.P., Hellmich, R.L., & Rice, L.G., (1999) *Plant Dis.* **83**, 130

(9)   Shepherd, G.S., Sydenham, E.W., Thiel, P.G., & Gelderblom, W.C.A., (1990) *J. Liq. Chromatogr.* **13,** 2077

(10)   Bakan, B., Melcion, D., Richard-Molard, D., & Cahagnier, B., (2002) *J. Agric. Food Chem.* **50**, 28

(11)   Visconti A, Solfrizzo M, & De Girolamo A., (2001) *J. AOAC Int.* **84**, 1828

(12)   Pietri, A. & Piva, G., (2000) in *Proceedings of the 6th International Feed Conference, Food Safety: Current Situation and Perspectives in the European Community.* Piva, G, Masoero F. (Eds), Piacenza, Italy 27-28 November, 2000. 226

(13)   Hammond, B.G., Campbell, K.W., Pilcher, C.D., Degooyer, T.A., Robinson, A.E.,McMillen, B.L., Spangler, S.M., Riordan, S.G., Rice, L.G., & Richard, J.L., (2004) *J. Agric. Food Chem.* **52,** 1390

(14)   de Castro, M.F., Shephard, G.S., Sewram, V., Vicente, E., Mendonca, T.A., & Jordan, A.C.,(2004) *Food Addit. Contam.* **21,** 693

(15)   Marasas W.F., Riley R.T., Hendricks K.A., Stevens V.L., Sadler T.W., Gelineau-van Waes J., Missmer S.A., Cabrera J., Torres O., Gelderblom W.C., Allegood J., Martinez C., Maddox J., Miller J.D., Starr L., Sullards M.C., Roman A.V., Voss K.A., Wang E., & Merrill A.H. Jr., (2004) *J. Nutr.* **134**, 711

(16) Committee on Genetics, American Academy of Pediatrics (1999) *Pediatrics* **104**, 325

(17) Bankole, S.A. & Mabekoje, O.O., (2004) *Food Addit. Contam.* **21,** 251

(18) Carlson, D.B., Williams, D.E., Spitsbergen, J.M., Ross, P.F., Bacon, C.W., Meredith, F.I., & Riley, R.T., (2001) *Toxicol Appl. Pharmacol.* **172**, 29

(19) Gelderblom, W.C.A, Marasas, W.F.O., Lebepe-Mazur, S., Swanevelder, S., Vessey, C.J.,& de la Hall, P., (2002) *Toxicology* **171**, 161

19. THE HAACP CONCEPT AND MYCOTOXINS

Douglas L. Park[1], Henry Njapau[2] and Ezzeddine Boutrif[3]

[1]Park and Associates
Cabot, Arkansas, USA

[2]Center for Food Safety and Applied Nutrition
Food and Drug Administration
College Park, Maryland, USA

[3]Food and Nutrition Division
Food and Agriculture Organisation
Rome, Italy

Notice: This chapter is based on an article entitled "Minimizing Risks Posed by Mycotoxins Utilizing the HACCP Concept" by Park, D.L., Njapau, H. and Boutrif, E., 1999. In: *Preventing Mycotoxin Contamination,* Food Nutrition, and Agriculture Publication Number 23, Albert, J.L. (Ed.) Food and Agriculture Organization, Rome, Italy, 49-54. Portions of the original publication are reproduced here with permission.

The potential of unavoidable naturally occurring toxicants contaminating agricultural commodities poses a unique challenge to food safety surveillance and management. As highlighted in a comprehensive report by the United Nations Food and Agriculture Organization (FAO), at least 25 percent of the world's food crops are contaminated with mycotoxins, at a time when the production of agricultural commodities is barely sustaining the increasing population (1). Worldwide, high risk commodities such as corn, peanuts, copra, palm nuts and oilseed cake, make up close to 100 million MT -20 million MT of which come from the developing countries (2). The destruction of contaminated products or their diversion to non-human uses is not always practical and could seriously compromise the world food supply. Efforts to limit mycotoxins in human foods and animal feedstuffs are based on two major concerns: the adverse effects of mycotoxin-contaminated crops or feeds on human or animal health and productivity; and potential residues of mycotoxin or toxic metabolites in edible animal products. The adverse effects of mycotoxins on human health were highlighted by the 2004 outbreak of aflatoxicosis in Kenya that resulted in 125 fatalities (3).

The concept of Hazard Analysis Critical Control Point (HACCP) has been used by the food industry for several decades to improve product quality and safety. HACCP is a system of food product potential contaminant control based on the systematic identification and assessment of a hazard or hazards in food and the definition of the means to control the hazard. It is a preventive, rather than a reactive, tool that places the maintenance and protection of the food supply from microbial, chemical and physical hazards as well as quality, into the hands of food management systems. The HACCP system is designed to minimize the risk of food safety hazards by identifying the hazards, establishing controls and monitoring these controls (4-6). The HACCP concept can be applied to the management of likely adverse health effects resulting from exposure to mycotoxins that would result in an adequate, wholesome and safe food supply. An effective HACCP-based integrated mycotoxin management program can be set up in both developed and economically challenged regions of the world. Each country or region should consider such factors as the climate, farming systems, preharvest and postharvest technologies, public health significance of the contaminant, producer and processor compliance, availability of analytical resources, and last but not least, the economy (6, 7).

The best method for controlling mycotoxin contamination is the prevention of mycotoxin formation through pre-harvest management. However, should contamination occur, the hazards associated with the toxins must be managed through postharvest procedures, if the product is to be used for food and feed purposes. In an ideal integrated mycotoxin management system, mycotoxin hazards would be minimized at every phase of production, harvesting, processing, and distribution (7, 8).

HACCP PROGRAM FOR MYCOTOXIN CONTAMINATION

The initial HACCP concept was used by the food industry to control product quality and microbial contamination. In the current terminology of HACCP systems, "hazard" refers to conditions or contaminants in foods that can cause illness or injury. It does not refer to certain quality or other undesirable conditions or contaminants such as insects, hair, filth and spoilage, which would be considered in the context of a broader quality assurance system. The HACCP concept is built on seven principles and actions:

- Conduct hazard analysis and identify preventive measures
- Identify critical control points (CCPs)
- Establish critical limits
- Monitor each CCP
- Establish corrective action in the event of a deviation from a critical limit
- Establish record keeping
- Establish verification procedures

The development and application of HACCP programs are complex matters, and not all countries have the required resources and/or technical expertise and experience to establish highly effective integrated mycotoxin management systems that are based on the HACCP approach. Given the importance of HACCP in food safety programs, FAO has given high priority to the provision of training to professionals in developing countries on the HACCP approach and its application (FAO, 1995).

Integrated mycotoxin management systems based on the HACCP approach including those set up in economically challenged regions, must, of course, consider hazards at all stages of production, handling and processing. Furthermore, a prerequisite for the development of HACCP programs is observance of good manufacturing practices (GMP) (9).

In the field, mycotoxin contamination is primarily the result of environmental conditions such as ambient temperature, precipitation, relative humidity, moisture of the product and its susceptibility, and the mold inocula that occur naturally throughout the world (*Aspergillus, Penicillium, Fusaria*, etc). It has now universally been acknowledged that mycotoxin formation may also occur at various stages of processing. Based on findings in the literature regarding mycotoxins, control can be effected during preharvest, harvest, and postharvest phases which, for the purposes of this article, will include storage and all forms of processing. Regardless of economic resources available, it is important that these points in the food handling systems are identified, understood and well managed. In an integrated mycotoxin management incorporating the HACCP concept, each identified and appropriately managed phase will help prevent the risk of contamination of a product.

The prevention and control of mycotoxin contamination to reduce qualitative and quantitative losses in food and agricultural products were integrated into many of FAO's Prevention of Food Losses projects. Of about 200 projects within this program, over 50 included components for mycotoxin control (1).

Farm-to-Table Mycotoxin Control

Drought, insect infestation, fungal inoculum, and delayed harvesting are important external factors that can result in significant levels of mycotoxins in food crops in the field. Some of these factors are environmental and humans have minimal control over them. However, good crop husbandry practices, such as crop rotation, irrigation, timed planting and harvesting, and the use of pesticides are preventative actions that reduce mycotoxin contamination of field crops. Numerous studies have shown that insect infestation can serve as a vector for mold infection in commodities that are susceptible to mycotoxin formation. Reduction of insect infestation is, therefore, critical for pre-harvest mycotoxin control. The main objective is to prevent mycotoxin formation at this phase of food production. Field crops should be harvested in a timely manner. If harvested at high moisture content, the moisture should be quickly reduced to levels at which mycotoxin formation would not occur. Damage to food commodities, i.e., kernels, nuts, seeds, etc., must be kept to a minimum during this phase. Mechanical damage to commodities facilitates fungal infection and consequently mycotoxin contamination.

Prevention through preharvest management is the best method for controlling mycotoxin contamination; however, should the contamination occur or persist after this phase, the hazards associated with the toxins must be managed through postharvest procedures if the product is to be used for food and feed purposes. In the postharvest phase, storage and processing are the major areas where contamination can be prevented. Processing can involve the removal of parts of the commodity, which may make it more susceptible to mold infection.

Storage, whether in the home, on the farm, at the manufacturing premises or in the grocery store, is the most critical postharvest phase in food handling. Factors that promote mold infection and subsequent mycotoxin formation must also be controlled. An inappropriate storage facility, improper packaging and/or the state of the food product can cause mycotoxin contamination during storage. The moisture levels of susceptible commodities, such as cereal kernels, nuts, seeds, etc., should be reduced to levels that do not support mold growth and mycotoxin formation. An accumulation of moisture and heat and/or physical damage to the product enhances fungal invasion, leading to the occurrence of mycotoxins. Stored products should not be exposed to environmental conditions, such as moisture, that promote mold growth. Neither should storage pests be allowed to be present in large enough numbers to cause significant physical damage to the product. Appropriate packaging is often a successful way of excluding insect pests and, where deficits in packaging are likely, general hygiene and the use of pesticides could help minimize contamination. When operating limits are violated, operators of storage facilities need to adjust the process followed. Under such a scheme mycotoxin contamination is detected before it is unmanageable, and the situation is rectified.

After the preharvest and harvesting phases, a commodity may undergo various changes during processing. This is another stage at which mycotoxins could be intentionally eliminated or their formation unintentionally enhanced. The nature of the processing procedure may increase the likelihood of clean commodities becoming colonized by molds and the subsequent production of toxins. Decreasing contamination at this phase ensures that the product reaching the consumer will have minimal likelihood of the hazards associated with mycotoxin contamination of food.

Once a contaminated product has been identified at a processing facility, clean-up and separation are the first alternatives of control. For example, electronic sorting and hand-picking to remove damaged, immature or mold-infected kernels, grains or nuts can remove a significant portion of the aflatoxins in shelled peanuts. The separation of grain into fractions by milling, followed by the elimination of the toxic portions is another decontamination strategy. Complete separation of all contaminated particles may not, however, be achieved since the toxin can diffuse into the interior of the kernel. Other procedures must, therefore, be used to manage the contamination in the final product.

Thermal inactivation is a good alternative for products that are usually heat processed. Fumonisin and ochratoxin levels have been shown to be lower in thermally processed corn and wheat products. On the other hand, aflatoxins and deoxynivalenol are resistant to thermal inactivation and are not destroyed completely by boiling water or a variety of food and feed processing procedures. Thermal inactivation for use at a CCP during processing should be evaluated for the conditions of the particular process and the subsequent fraction(s) containing mycotoxin residues identified.

Other Potential Control Processes

Exploring the application of other processes to control mycotoxins could widen the opportunities for risk management. Where physical separation and thermal inactivation are inappropriate, other techniques may be employed, as long as the end product is acceptable to the consumer. For instance, ammoniation has been successfully employed to reduce aflatoxin contamination in corn, peanuts, cottonseed and peanut and cottonseed meals. This procedure when applied to aflatoxin-contaminated dairy ration ingredients can reduce potential aflatoxin residues in milk to non-detectable levels. By including hydrogen peroxide and sodium bicarbonate in the nixtamalization procedure (a traditional alkaline heat treatment of corn used in the manufacture of tortillas), its efficacy against fumonisin and aflatoxin toxicity is increased.

In such industries as oil refining, the use of adsorbent materials is part of normal processing operations. A variety of adsorbent materials, i.e., activated carbon and clays, have been shown to bind aflatoxins in aqueous solutions, and certain aluminosilicates have been reported to bind aflatoxins in peanut oil and animal feeds. Phyllosilicate clay has been shown to prevent acute aflatoxicosis in farm animals and decrease the levels of aflatoxin M_1 residues in milk. Activated charcoal has also proved to be effective in reducing the patulin in naturally contaminated fruit juices. However, since some adsorbent materials may pose a greater risk than benefit, care must be taken when choosing these products. Table 19.1 shows stages at which the HACCP program could be applied to mycotoxin control.

Table 19.1. Possible stages in application of the HACCP principle to agricultural commodities, food products and animal feedstuffs

Critical Control Point	Commodity	Hazard	Corrective Action
Pre-harvest	Cereal grains, oil seeds, nuts, fruits	Mold infestation with subsequent mycotoxin formation	• Utilize crop resistant varieties • Enforce effective insect control programs • Maintain adequate irrigation schedules • Perform good tillage, crop rotation, weed control practices, etc.
Harvesting	Cereal grains, oil seeds, nuts, fruits	Increase in mycotoxin formation	• Harvest at appropriate time • Maintain at lower temperature, if possible • Remove extraneous material • Dry rapidly to below 10% moisture
Post-harvest and storage	Cereal grains, oil seeds, nuts, fruits	Increase and/or occurrence of mycotoxin	• Protect stored product from moisture, insects, environmental factors, etc. • Store product on dry clean surface
Post-harvest, processing and manufacturing	Cereal grains, oil seeds, nuts, fruits	Mycotoxin carryover or contamination	• Testing of all ingredients added • Monitor processing/manufacturing operation to maintain high quality product • Follow good manufacturing practices
Animal feeding	Dairy, meet and poultry products	Transfer of mycotoxin to dairy products, meand and poultry products	• Monitor mycotoxin, levels in feed ingredients • Test products for mycotoxin residues

Reproduced from: Park, D.L., Njapau, H. Boutrif, E., 1999.

ASSOCIATED PROGRAMS IN MYCOTOXIN CONTROL

Effective integrated mycotoxin management programs not only cover prevention of mycotoxin formation in agricultural products or their detoxification/decontamination, but also involve: routine surveillance; regulatory measures to control the flow of mycotoxin- contaminated material in national and international trade; and information, education and communication activities.

FAO has been widely involved in reviewing national food control systems in its developing member countries, in many instances focusing on problems of mycotoxin control. Such reviews involve the identification of weaknesses in the food control infrastructure -i.e. the food control administration, inspection and analytical capacity, regulatory issues, etc. After the review process, FAO is frequently involved in the implementation of technical assistance projects to address existing problems.

FAO has played an important role in the dissemination of technical literature to its member countries in support of their efforts in the area of mycotoxin control. Several publications and training aids on various aspects of mycotoxin control have been prepared and distributed. The publications cover a variety of topics, such as methods of sampling and analysis; a training syllabus for use in short- term courses on aflatoxin analysis; a compilation of mycotoxin regulations; and directories of mycotoxin prevention and control institutions in selected regions. The latter are examples of FAO's efforts to facilitate regional networking so as to optimize the use of scarce financial, human and technical resources in the implementation and upgrading of mycotoxin management programs. Such networking is also promoted through FAO's regional mycotoxin projects and programs. The focus of the Workshop on Mycotoxins hosted by the U.S. Food and Drug Administration, College Park, Maryland, in 2002 was to provide tools to assist economically challenged regions of the world to have effective mycotoxin control programs.

Establishment of Regulatory Limits

Hazard analysis and CCP systems must be built on existing or simultaneously established food safety programs in each country. Implementation of HACCP cannot be expected to succeed in the absence of regulatory activities. Regulatory limits or standards provide a benchmark against which the effectiveness of food safety programs can be tested. Regulatory limits are law, violation of which has legal consequences. Regulatory limits and/or standards must reflect local regulations as well as potential trade partners.

FAO has supported the compilation of information, at regional and global levels, on maximum tolerable levels for mycotoxins in food and feed. The last global compilation dates from 1995 and contains data from 90 countries on mycotoxin regulations, tolerance levels and methods of sampling and analysis. It has been published and widely distributed (11).

The importance of the development of internationally harmonized regulatory mycotoxin control measures that protect public health and promote fair trade at the international level cannot be overemphasized. It is of particular importance in view of the World Trade Organization (WTO) Agreements on Sanitary and Phytosanitary Measures (SPS) and Technical Barriers to Trade (TBT). These agreements call for greater harmonization and transparency in the establishment of food regulations that are meant to facilitate trade without compromising consumer protection. The SPS Agreement states that measures conforming to international Codex standards, guidelines or other recommendations are deemed to be appropriate, necessary and non-discriminatory. WTO's recognition of Codex standards, guidelines and other recommendations as benchmarks for food safety is undoubtedly related to the role of science in the Codex process. The Joint FAO/WHO Expert Committee on Food Additives (JECFA) plays an important role in the elaboration of Codex standards and guidelines related to mycotoxin contamination by providing evaluations based on sound scientific and risk assessment principles.

Implementation of Surveillance and Monitoring Programs

Surveillance and monitoring activities fall within HACCP principles 4 to 7. In the absence of regulation and surveillance, voluntary compliance with any system may not be wholly achieved. Announcing the seafood HACCP regulation in 1995, David Kessler, the US Food and Drug Administration Commissioner at the time, stated "Our safety inspections should focus on preventing problems rather than chasing the horses after they are out of the barn". Safety inspections are the responsibility of government regulatory agencies that ensure the adequacy of industry food safety programs. Inspections to uncover violations are based on set limits and standards. In order to set up a monitoring and management program, the following data and policy decisions must be acquired and made based on available resources:

- Identify the mycotoxin(s) and the products or commodities that are to be included in the program

- Set up a system of inspection and sample collection

- Set up a sampling plan

- Establish a policy guide for end use of the products, for example:

  o proceed into market channels "as is";

  o use as designated animal feed, e.g. dairy, feed lot, finishing, starter;

  o divert to decontamination procedures or lower-risk uses.

Once a regulatory limit has been set, monitoring programs play an important role in determining compliance. For mycotoxin contamination, it is important to consider adequate random sampling techniques that consider the existence of "hot spots" or highly contaminated portions of the product. A well-designed sampling plan and validated methodology will provide, within limitations, the concentrations of specific analyte(s) for a specified lot of material. It is important that the analyst is competent to conduct the method. The analytical result is of no value if the sample collected and prepared for analysis does not represent the lot and conceals or overexpresses violations of critical limits. Care must therefore be taken to assure that proper procedures are followed.

A good example of management through a monitoring program is the aflatoxin control program established by the State of Arizona in the United States. In 1978, almost 414, 000 kg of milk were discarded because of high aflatoxin M_1 levels. As a result of this huge commercial loss, the state instituted a program to monitor aflatoxin levels in whole cottonseed and cottonseed products used in dairy rations at processing and dairy farm points. All cottonseed produced in the state is tested for aflatoxin content. The maximum size of the lots tested is 100 tons, and the testing is conducted in state-certified laboratories. The end use of the product is dictated by the aflatoxin levels that are found. Cottonseed lots testing over 20 µg of aflatoxin per kilogram are usually treated with ammonia to reduce these levels, and then re-tested. The use of this program has kept Arizona's milk supply safe from aflatoxin residues. The same concept can also be applied to other commodities.

The establishment of monitoring and surveillance programs for mycotoxins requires suitably equipped laboratories, well-trained staff for both analytical and inspection activities, reliable analysis and sampling methods, and application of analytical quality assurance programs. It can be difficult for developing regions to meet these requirements. Under these circumstances the surveillance and monitoring programs may have to be modified in line with available resources. Specific FAO projects in a number of countries have addressed these issues in providing technical assistance in the area of surveillance of mycotoxin contamination. Although aflatoxin was the first priority in most of the projects, they all also gave some attention to the surveillance of other mycotoxins. Existing systems for monitoring food contaminants, including various mycotoxins have been studied and strengthened in Asian countries such as Bhutan, China, India, Indonesia, Nepal, Pakistan, The Philippines, Sri Lanka, Thailand and Vanuatu; Latin American countries such as Chile, Cuba, Guatemala and Uruguay; and in African countries such as Malawi, Rwanda, the United Republic of Tanzania and several West African States (1).

Analytical quality assurance studies were carried out at regional level in Latin America and Asia. Results highlighted the need for continuing such exercises and increasing the number of participating laboratories.

As a means of building sustainable national capacity, training has been a major component of FAO's assistance to developing countries in improving mycotoxin control. A long-term, international training program was carried out during the 1980s in collaboration with the United Nations Environment Program (UNEP) and what was then the Union of Socialist Soviet Republics (USSR). Other activities included local and regional courses for laboratory staff and practical demonstrations on field detection, identification and analysis of various mycotoxins. In Asia, a training network was implemented to provide training in methods of analysis and sampling of various mycotoxins, including guidance on policy issues. Regional workshops on mycotoxin analysis were held in Senegal, Botswana and various Latin American countries.

SUMMARY AND CONCLUSIONS

Much has been accomplished at national, regional and international levels regarding mycotoxin prevention and control, but much still remains to be done. It is now widely recognized that food safety programs should be based on the strict observance of good agricultural, processing and handling practices, including the application of HACCP concepts; governments should therefore upgrade their mycotoxin management programs to include the HACCP principles. HACCP-based mycotoxin management programs must involve control and surveillance at all stages of production and post-production, as there are several factors at pre-harvest, harvest and post-harvest stages that are implicated in mycotoxin contamination of crops. FAO has been active in providing assistance to its member countries in various aspects of mycotoxin management including mycotoxin prevention and control, routine surveillance and regulatory matters. Minimizing the risks posed by mycotoxins through applying good agricultural, processing and handling practices and utilizing the HACCP concept continues to be priority for FAO.

Since mycotoxin contamination of susceptible commodities can occur as a result of environmental conditions in the field and improper harvesting, storage and processing operations, comprehensive surveillance and control programs must be established. Hazard Analysis Critical Control Point (HACCP) programs have been useful in the detection and management of the risks associated with potential contamination of food products with pathogenic microorganisms and chemical toxicants. Food safety programs routinely use information about the factors leading to contamination to establish preventive and control procedures, thus providing the consumer with a safe, wholesome food supply. When an effective HACCP program for mycotoxins has been established, key elements are identified that can be used or modified to detect and reduce mycotoxin formation in field and storage environments. HACCP programs can be designed to apply to large commodity growing enterprises, industrial storage and processing operations, and food distribution and retail operations as well as circumstances there are limited economic resources. The HACCP elements would include limiting insect infestation and moisture levels in the commodities. Specific processing and decontamination procedures can play a role in reducing mycotoxin levels through the physical separation of damaged, immature and mold-infested kernels, grains or nuts, and the physical and chemical inactivation and/or removal of the toxin. The development and application of HACCP-based food monitoring and safety programs require expertise in a range of fields. FAO has been active in providing technical assistance to its member countries, helping to build a national capacity to implement and maintain effective HACCP-based mycotoxin management programs.

REFERENCES

(1)   Boutrif, E. & Canet, C. (1998) *Revue Méd. Vét.*, **149,** 668

(2)   FAO (1996) *Basic Facts on the World Cereal Situation.* Food Outlook, 5/6, Food and Agriculture Organization, Rome, Italy

(3)   Lewis, L., Onsongo, M., Njapau, H., Schurz-Rogers, H., Luber, G., Kieszak, S., Nyamongo, J., Backer, L., Dahiye, A.M., Misore, A., DeCock, K., Rubin, C., and the Kenya Aflatoxicosis Investigation Group (2005). *Environ. Health Perspect.* **113,** 1763

(4)   FAO. (1995) *The Use of Hazard Analysis Critical Control Point (HACCP) Principles in Food Control.* FAO Food and Nutrition Paper No. 58. Food and Agriculture Organization, Rome, Italy

(5)   Gamboa, D.E. (1998) *Dairy Food Environ. Sanit.*, **18,** 288

(6)   FAO. (2001) *Manual on the Application of the HACCP System in Mycotoxin Prevention and Control,* Food and Nutrition Paper No. 73, Food and Agriculture Organization, Rome, Italy

(7)   FAO. (1979) *Recommended Practices for the Prevention of Mycotoxins in Food, Feed and their Products.* FAO Food and Nutrition Paper No. 10. Food and Agriculture Organization, Rome, Italy

(8)   Lopez-Garcia, R. & Park, D.L. (1998) *in Mycotoxins in Agriculture and Food Safety,* Bhatnagar, D & Sinha, K.K (Eds)., Marcel Dekker, New York.

(9)   Sperber, W.H., Stevenson, K.E., Bernard, D.T., Deibel, K.E., Moberg, L.J., Hontz, L.R. & Scott, V.N. (1998) *Dairy Food Environ. Sanit.,* **18,** 418.

(10)   Park, D.L., Lee, L.S., Price, R.L., & Pohland, A.E. (1988) *J. AOAC Int.,* **71,** 685.

(11)   FAO. (1997) *Worldwide Regulations for Mycotoxins 1995 -A Compendium.* FAO Food and Nutrition Paper No. 64., Food and Agriculture Organization, Rome, Italy

(12)   FAO. (1993) *Sampling Plans for Aflatoxin Analysis in Peanuts and Corn. A Technical Consultation.* FAO Food and Nutrition Paper No. 55., Food and Agriculture Organization, Rome, Italy

20. INTERNATIONAL RISK ASSESSMENT OF MYCOTOXINS

Gerald G. Moy

Food Safety Program
World Health Organization
Geneva, Switzerland

Since its inception in 1949, the World Health Organization (WHO) has had a major interest in food safety as part of its overall health mandate. For example, in 1953 WHO published two monographs concerning important food safety issues at that time: one on milk pasteurization (1) and another on pesticides (2). WHO and its member states have long recognized the importance of food safety to health and development (3). More recently, in May 2000, the WHO World Health Assembly adopted a resolution that emphasized food safety as an essential public health function and called on all member states to take stronger measures to ensure the safety of their food supplies from both chemical and biological hazards.

The old adage "we are what we eat" suggests that people have generally recognized the relationship between good food and good health. Indeed, laws enacted by many ancient civilizations from around the world confirm that the earliest governments were concerned about the quality and safety of the food supplies. Some religious requirements are based on sound hygienic principles that remain valid today. Unfortunately, the correlation between the consumption of moldy food and mycotoxicoses has been generally overlooked. Nonetheless, the historically recent recognition of the potential hazards posed by mycotoxins in food has brought into play well-established scientific and management mechanisms, which have evolved to ensure the safety of the food supply.

Ensuring the safety of the food supply has become one of the most important and increasingly complex functions carried out by governments. Risk analysis has become the preferred approach for addressing current and emerging food safety problems, especially for chemicals such as mycotoxins. The evaluative framework offered by risk analysis not only permits an estimation of human risk but also provides a means of organizing data and allocating responsibilities. Risk analysis permits a transparent and relatively uniform methodology to assess risks posed by specific mycotoxins, to manage such risks, and to facilitate communication between risk assessors and risk managers as well as all stakeholders. It is not surprising that risk analysis has been adopted by many national food safety authorities as well as the Codex Alimentarius Commission as the preferred approach for addressing health risks posed by chemical, microbiological, and physical hazards in the food supply. Although risk management and communication are important, the foundation of risk analysis is sound scientific risk assessment. This paper examines the use of risk assessment by international organizations, in particular WHO, in addressing the health risks posed by mycotoxins in food.

CONTEXT OF INTERNATIONAL RISK ASSESSMENT

With establishment of the World Trade Organization (WTO) in 1995, the reduction of tariffs and subsidies for agricultural commodities negotiated under the Uruguay Round of Multilateral Trade Negotiations became a reality. However, to avoid erecting nontariff barriers to trade on the basis of unjustifiable health and safety requirements, the Agreement on the Application of Sanitary and Phytosanitary Measures (the SPS Agreement) was included as a side agreement. In addition to other disciplines, the SPS Agreement requires WTO member-countries to base their health and safety requirements for food on sound scientific risk assessment, taking into account the methods developed by relevant international bodies. For food safety, this body is the Codex Alimentarius

Commission, an intergovernmental body established by WHO and the Food and Agriculture Organization of the United Nations (FAO) in 1963 with the purpose of *inter alia* "protecting the health of consumers and ensuring fair practices in the food trade" (4). Codex, now with 167 member countries, has developed a considerable body of standards, guidelines, and other recommendations that, if adopted by a country, automatically ensure that country is in compliance with the SPS Agreement.

Codex receives its risk assessment advice from various expert bodies convened by WHO and FAO and at times with the participation of other international organizations, such as the Pan American Health Organization (PAHO), the International Atomic Energy Agency (IAEA), and the World Animal Health Organization (WAHO). To ensure consistent and transparent application of risk assessment across all Codex activities, a Joint FAO/WHO Expert Consultation on the Application of Risk Analysis to Food Standards Issues was convened in Geneva in March 1995 (5). The consultation made a number of recommendations for improving Codex risk assessment procedures, including proposing definitions for risk analysis. The risk analysis definitions with some modifications were subsequently adopted by the 22nd Session of the Codex Alimentarius Commission in June 1997; they provide a framework for risk analysis—in particular, risk assessment (4).

RISK ASSESSMENT OF MYCOTOXINS BY INTERNATIONAL AGENCIES

Sound scientific risk assessments are routinely provided by WHO and FAO expert bodies to support the work of Codex. These assessments are likely to become increasingly important as the basis for arbitration involving health-related trade disputes among countries. At present, Codex has developed or is in the process of developing recommendations to reduce levels of mycotoxins in food, including establishing maximum limits for levels of mycotoxins in commodities. Therefore, it would be instructive to examine the international risk assessment model, which has become the basis for all Codex risk management. In principle, all WTO member countries should also take this model into account when conducting their own risk assessments. The risk assessment process generally follows the chronology discussed below, but the process is usually more interactive between risk assessors and managers. For mycotoxins, it is also more iterative as new information relevant to risk assessment becomes available.

Hazard Identification

With identification of aflatoxin as the cause of Turkey X disease in the 1960s, many countries recognized the potential human health risks posed by mycotoxins. Several other mycotoxins were later identified as potential hazards, mainly as a result of outbreaks of disease in animals and, at times, in humans. However, it may be said that the human health effects caused by exposure to mycotoxins in food are complex and poorly understood. Nonetheless, available human epidemiology data suggest that mycotoxins may be responsible for a range of human diseases resulting from acute and chronic exposure (6). Diseases thought to be associated with acute exposure to mycotoxins include ergotism (ergot alkaloids), alimentary toxic aleukia (trichothecenes), yellow rice disease (citrinin), sugarcane poisoning (3-nitroprionic acid), acute hepatitis (aflatoxins), and kwashiorkor (aflatoxins). Diseases with possible chronic mycotoxin exposure etiologies include esophageal cancer (fumonisin), Indian childhood cirrhosis (aflatoxin), primary hepatocellular carcinoma (aflatoxin), and Balkan endemic nephropathy (ochratoxin A).

Within Codex, hazard identification begins with a draft position paper, which is prepared by a country or groups of countries for presentation to the Codex Committee on Food Additives and Contaminants (CCFAC), the Codex subsidiary body responsible for contaminants, including mycotoxins. Priorities are ranked based on several criteria, including the toxicity of the substance, estimated dietary exposure, potential trade problems, and availability of adequate information. For many mycotoxins, the main impediment has been the paucity of information upon which to conduct a risk assessment. Once a substance is placed on the CCFAC priority list, it is referred to the Joint FAO/WHO Expert Committee on Food Additives (JECFA) for evaluation. It should be noted that scheduling a substance for JECFA review depends on other priorities submitted by WHO and FAO member countries and on the availability of space on the JECFA agenda.

Hazard Characterization

Since 1956, JECFA has been engaged in assembling and evaluating scientific data on food additives and making recommendations on safe levels of use. In 1972, the scope of the evaluations was extended to include contaminants in food. One of the major uses of JECFA recommendations is to serve as the scientific basis for developing Codex standards, guidelines, and other recommendations. In general terms, the purpose and functions of JECFA include (a) reviewing the latest knowledge and expert information and making it available to FAO, WHO, and their member countries; (b) formulating technical recommendations; and (c) making recommendations designed to initiate, stimulate, and coordinate the research necessary to reach conclusions about the toxicological acceptance or otherwise of the presence of a substance in food (7).

In the past, insufficient data for many mycotoxins precluded their evaluation by JECFA. More recently, however, JECFA has provided a number of useful risk assessments, which offer guidance for risk managers in countries and in Codex. The hazard characterization component usually has been provided in the form of a provisional tolerable weekly intake (PTWI) or a provisional maximum tolerable daily intake (PMTDI). These are presented for several mycotoxins in Table 20.1.

The WHO International Agency for Research on Cancer (IARC) has undertaken extensive studies to assess the carcinogenicity of the various mycotoxins. The IARC has concluded that there is sufficient evidence for the carcinogenicity of naturally occurring mixtures of aflatoxins in humans and that aflatoxin M_1 is possibly carcinogenic to humans (13). The IARC has also reviewed a number of other mycotoxins, which are summarized in Table 20.2. Several mycotoxins have been classified by IARC as possibly carcinogenic to humans (category 2B). However, IARC has made no estimate of the potency of aflatoxin B_1 as a human carcinogen.

Table 20.1. Summary of JECFA hazard characterizations of mycotoxins

Mycotoxin	PMTDI (ng/kg body weight)	JECFA meeting
Deoxynivalenol	1000	56th (8)
Fumonisin B_{1-3}	2000	56th (8)
Ochratoxin A	100 (PTWI)	35th (9), 37th (10), 44th (11), 56th (8)
Patulin	400	35th (9), 44th (11), 56th (8)
T-2 and HT-2 toxins	60	56th (8)
Zearalenone	500	53rd (12)

Table 20.2. Summary of IARC evaluations of mycotoxins (13-16)

Agent	Degree of evidence of carinogenicity		Overall evaluation of carcinogenicity to humans
	Human	Animal	1
Aflatoxins, naturally occurring mixture	S	S	
Aflatoxin B1	S	S	
Aflatoxin B2		L	
Aflatoxin G1		S	
Aflatoxin G2		I	
Aflatoxin M1	I	S	2B
Citrinin	I-NDA	L	3
Cyclochlorotin	I-NDS	L	3
Griseofulvin	I-NDA	S	2B
Luteoskyrin	I-NDA	L	3
Ochratoxin A	I	S	2B
Patulin	I-NDA	I	3
Penicillic acid	I-NDA	L	3
Rugulosin	I-NDA	I	3
Sterigmatocytin	I-NDA	S	2B

Toxins derived from:			
Fusarium graminearum, *Fusarium culmorum, and* *Fusarium crookwellense*	I-NDA		3
Zearalenone		L	
Deoxynivalenol		I	
Nivalenol		I	
Fusarenone X		I	
Toxins derived from:			
Fusarium moniliforme	I	S	2B
Fumonisin B1		L	
Fumonisin B2		I	
Fusarin C		L	
Toxins derived from:			
Fusarium sporotrichioides	I-NDA		3
T-2 toxin		L	

Evidence: S = sufficient; L = limited evidence; I = inadequate; I-NDA = inadequate, no data available. Overall evaluation: I = crcinogenic to humans, 2B = possibly carcinogenic to humans, 3 = not classifiable as to its carcinogenicity to humans.

With regard to aflatoxins, the Forty-ninth JECFA, after considering several epidemiology studies of the carcinogenic potency of aflatoxin B_1, estimated the potency of aflatoxins B and G to be 0.01 cancer per year per 100, 000 population per ng of aflatoxin per kg of body weight. The potency was estimated to be about 30 times higher for carriers of hepatitis B virus, which is also a risk factor for primary hepatocellular carcinoma (17). The Fifty-sixth JECFA further estimated the carcinogenic potential of aflatoxin M_1, which is a metabolite of aflatoxin B_1 present in cow's milk, to be about one-tenth that of aflatoxin B_1 (8). The Fifty-sixth JECFA was devoted solely to the topic of mycotoxins and the monograph prepared by this meeting is recommended for anyone interested in the topic of risk assessment of mycotoxins (18).

Exposure Assessment

At the international level, the need for a global assessment of chemicals in air, water, and food led to the establishment in 1976 of the Global Environment Monitoring System—Food Contamination Monitoring and Assessment Program under the auspices of WHO, FAO, and the United Nations Environment Program. Commonly known as GEMS/Food, the program is now implemented by WHO through its network of participating institutions located in nearly 70 countries around the world. GEMS/Food is intended to inform governments, the Codex Alimentarius Commission, and other relevant institutions, as well as the public, on levels and trends of contaminants in food, their contribution to total human exposure, and significance with regard to public health and trade. Nineteen contaminants are currently included in the GEMS/Food Core List, including aflatoxins. Several other mycotoxins, including ochratoxin A, patulin, and fumonisin B_1, appear on the GEMS/Food Intermediate and Comprehensive Lists. Exposure assessment of food contaminants is especially dependent on adequate and reliable monitoring programs to provide accurate assessments of levels of contaminants in food.

The GEMS/Food database is periodically evaluated to assess levels and trends in food contamination. The most recent assessment of aflatoxin data from 30 countries conducted by GEMS/Food concluded that many countries, especially developed nations, were taking effective steps to limit aflatoxins in food and feeds. However, the instances in which relatively high levels were reported emphasized the necessity to intensify efforts to control aflatoxin levels in the global food supply, especially in corn, peanuts, tree nuts, pulses, and animal feeds. This was particularly true for many of the developing countries, where the climate is conducive to the growth of molds and resources are limited for preventing the formation of mycotoxins in commodities and for controlling their levels in the food supply (19).

GEMS/Food has been involved in developing methods for assessing exposure to chemicals in food. Beginning in 1989, GEMS/Food provided the exposure assessment for pesticides considered by the Codex

Committee on Pesticide Residues. More recently, these assessments have been integrated into the risk assessment performed by the Joint FAO/WHO Meetings on Pesticide Residues, the pesticide counterpart of JECFA. These assessments were based on food consumption provided by GEMS/Food for five regions of the world (20). The development of 13 so-called "consumption cluster" diets is being undertaken to improve accuracy and to better represent diet patterns. The method and use of GEMS/Food diets have been used by JECFA for assessing exposure to mycotoxins and other contaminants. However, exposure assessments must be tailored to the specific toxicological reference point, such as the PMTDI, PTWI, or potency.

Although JECFA has not established an acute reference dose for any mycotoxin, many mycotoxins clearly possess acute toxicity (e.g., vomitoxin), which requires alternative risk management strategies to protect public health. Procedures have been developed to conduct a short-term exposure assessment for pesticide residues, which may be applicable to mycotoxins.

Risk Characterization

Risk characterization is defined as the qualitative and/or quantitative estimation, including attendant uncertainties, of the probability of occurrence and severity of known or potential adverse health effects in a given population based on hazard identification, hazard characterization, and exposure assessment. The risk characterization of mycotoxins with PTWIs or PMTDIs is usually expressed as the estimated exposure as a percentage of tolerable intakes. If the tolerable intakes were not exceeded, this would be expected to represent a tolerable risk. However, it should be recognized that this risk estimate is qualitative, because the only options are tolerable and not tolerable.

On the other hand, the risk characterization for aflatoxins provides a quantitative assessment that has more useful aspects than estimating risks. Comparative risks might be determined for assessing the impact of risk management interventions, such as setting various maximum limits. For example, the various scenarios for controlling aflatoxins in corn and peanuts were considered, including a maximum limit at 10, 15, and 20 µg/kg as well as having no limit. The analysis demonstrated that there was a clear public health advantage in having some limit, but the actual differences among the various proposed limits did not significantly affect health outcomes (17).

SUMMARY

The hazards posed by mycotoxins in food present an enormous challenge to national governments and international agencies responsible for their control. Deficiencies in the risk assessment database and inconsistencies in risk assessment procedures may confound any international consensus on tolerable levels for mycotoxins. Therefore, it is essential that national governments adopt a consistent, science-based risk assessment approach to address problems posed by mycotoxins in food, particularly with regard to exposure assessments. In this regard, the Codex Alimentarius Commission should be used to mediate competing risk management viewpoints to ensure consumer protection while preventing the erection of unwarranted nontariff barriers to trade in food. A strengthened JECFA and improved information for hazard characterization and exposure assessment must support this.

WHO remains committed to elucidating the possible role of mycotoxins in chronic diseases, such as atherosclerosis, for which clear etiologies have not been established. In addition, WHO continues to expand and improve its global food contamination monitoring program for mycotoxins under the auspices of GEMS/Food and in collaboration with its member states.

In light of the potential risks posed by mycotoxins, it would be prudent to continue to explore methods to prevent, reduce, or eliminate their presence in food, as is recommended by JECFA. Global research efforts should be strengthened, especially in developing countries. Dissemination of information and establishment of contact points, such as WHO Collaborating Centers for Mycotoxins in Food, can facilitate this process. Financial and human resources for research into the problem of mycotoxins should be pooled to accelerate the most promising new developments. In this regard, sophisticated technologies being developed in countries like the United States should be transferred to developing countries so that these technologies can be tested on a worldwide basis.

REFERENCES

(1)   WHO (1953) *Milk Pasteurization*, WHO Monograph Series, No. 14, World Health Organization, Geneva, Switzerland

(2)   WHO (1953) *Toxic Hazards of Certain Pesticides*, WHO Monograph Series, No. 16, World Health Organization, Geneva, Switzerland

(3)   WHO (1984) *The Role of Food Safety in Health and Development*, Report of a Joint FAO/WHO Expert Committee on Food Safety, Technical Report Series, No. 705, World Health Organization, Geneva, Switzerland

(4)   FAO (2001) *Procedural Manual*, 12th Ed., Joint Food and Agriculture Organization of the United Nations/World Health Organization Codex Alimentarius Commission, Secretariat of the Joint FAO/WHO Food Standards Program, Food and Agriculture Organization of the United Nations, Rome, Italy

(5)   WHO (1995) *Report of the Joint FAO/WHO Expert Consultation on the Application of Risk Analysis to Food Standards Issues*, 13–17 March 1995, Geneva, Switzerland WHO/FNU/FOS/95.3, World Health Organization, Geneva, Switzerland

(6)   Wild, C.P., & Hall, A.J. (1996) in *The Mycota VI—Human and Animal Relationships*, D.H. Howard & J.D. Miller (Eds), Springer-Verlag, Berlin, Germany

(7)   WHO (1996) *Summary of Evaluations Performed by the Joint FAO/WHO Expert Committee on Food Additives (JECFA)*,

Food and Agriculture Organization of the United Nations, International Life Sciences Institute, International Program on Chemical Safety, World Health Organization, Geneva, Switzerland

(8)   WHO (2002) *Evaluation of Certain Food Additives and Contaminants*, Fifty-sixth Report of the Joint FAO/WHO Expert Committee on Food Additives, WHO Technical Report Series, No. 906, World Health Organization, Geneva, Switzerland

(9)   WHO (1990) *Evaluation of Certain Food Additives and Contaminants*, Thirty-fifth Report of the Joint FAO/WHO Expert Committee on Food Additives, WHO Technical Report Series, No. 789, World Health Organization, Geneva, Switzerland

(10)   WHO (1991) *Evaluation of Certain Food Additives and Contaminants*, Thirty-seventh Report of the Joint FAO/WHO Expert Committee on Food Additives, WHO Technical Report Series, No. 806, World Health Organization, Geneva, Switzerland

(11)   WHO (1995) *Evaluation of Certain Food Additives and Contaminants*, Forty-fourth Report of the Joint FAO/WHO Expert Committee on Food Additives, WHO Technical Report Series, No. 859, World Health Organization, Geneva, Switzerland

(12)   WHO (2000) *Evaluation of Certain Food Additives and Contaminants*, Fifty-third Report of the Joint FAO/WHO Expert Committee on Food Additives, WHO Technical Report Series, No. 896, World Health Organization, Geneva, Switzerland

(13)   IARC (1993) *Some Naturally Occurring Substances: Food Items and Constituents, Heterocyclic Aromatic Amines and Mycotoxins,*

IARC Monographs on the Evaluation of Carcinogenic Risks to Humans, Volume 56, International Agency for Research on Cancer, Lyon, France

(14)   IARC (1976) *Some Naturally Occurring Substances*, IARC Monographs on the Evaluation of Carcinogenic Risks to Humans, Volume 10, International Agency for Research on Cancer, Lyon, France

(15)   IARC (1983) *Some Food Additives, Feed Additives and Naturally Occurring Substances*, IARC Monographs on the Evaluation of Carcinogenic Risks to Humans, Volume 31, International Agency for Research on Cancer, Lyon, France

(16)   IARC (1986) *Some Naturally Occurring and Synthetic Food Components, Furocoumarins and Ultraviolet Radiation*, IARC Monographs on the Evaluation of Carcinogenic Risks to Humans, Volume 40, International Agency for Research on Cancer, Lyon, France

(17)   WHO (1999) *Evaluation of Certain Food Additives and Contaminants*, Forty-ninth Report of the Joint FAO/WHO Expert Committee on Food Additives, WHO Technical Report Series, No. 884, World Health Organization, Geneva, Switzerland

(18)   WHO (2001) *Safety Evaluation of Certain Mycotoxins in Food*, WHO Food Additive Series 47, International Program on Chemical Safety, World Health Organization, Geneva, Switzerland

(19)   UNEP (1988) *Assessment of Chemical Contaminants in Food*, Report on the Results of the UNEP/FAO/WHO Program on Health-Related Environmental Monitoring, United Nations Environment Program Monitoring and Assessment Research Centre, London, England

(20)   WHO (1998) *GEMS/Food Regional Diets*, Document WHO/FSF/FOS/98.3, World Health Organization, Geneva, Switzerland

21. MYCOTOXIN CONTROL: WORLDWIDE REGULATIONS

Hans P. van Egmond[1] Douglas L. Park[2] and Henry Njapau[2]

[1]National Institute of Public Health and the Environment
Laboratory for Residue Analysis
P.O. Box 1, 3720 BA
Bilthoven, The Netherlands

[2]Center for Food Safety and Applied Nutrition
Food and Drug Administration
5100 Paint Branch Parkway
College Park, Maryland 20740

The discovery of the aflatoxins greatly stimulated scientific interest in mycotoxins and mycotoxicoses. Since 1960, thousands of publications about different aspects of mycotoxins have appeared. Many (but not all) of the mycotoxins cause illnesses in humans and animals, particularly in tropical and subtropical developing countries. The significance of the health effects of mycotoxins has led many countries to establish limits and regulations in recent decades to safeguard the health of humans and animals as well as the economic interest of producers and traders. Mycotoxin control programs, including the establishment of regulatory limits, monitoring schemes, and decontamination procedures, can minimize the hazard of mycotoxins to consumers.

FACTORS INFLUENCING THE SETTING OF MYCOTOXIN REGULATIONS

The setting of mycotoxin regulations is a complex activity that involves many factors and interested parties. Considerations that enter into regulatory decisions for controlling aflatoxin levels in foods and animal feeds are presented in Chapter 1. They include both scientific and socioeconomic factors. Risk assessment and risk characterization provide a sound scientific basis for establishing mycotoxin regulatory programs; where risk characterization information is lacking or insufficient, regulations may be established based on existing circumstantial association of consumption of the toxins with ill health.

Adequacy of Methods

Legislation calls for a capability to ascertain the occurrence and possibly violative quantity of the regulated substance, so reliable analytical methods must be available. Tolerance levels that do not have a reasonable expectation of being measured are wasteful in the resources they utilize and may well condemn products that are perfectly fit for consumption (1). In addition to reliability, simplicity is required, as it influences the amount of data that will be generated and the practicality of the ultimate measures taken. The reliability of analytical data can be improved if methods that fulfill certain performance criteria are used. AOAC INTERNATIONAL and the European Standardization Committee (CEN) have a number of standardized methods of analysis available that have been validated in formal collaborative studies, and this number is gradually growing. The latest edition of *Official Methods of Analysis* of the AOAC INTERNATIONAL (2) contains about 40 validated methods for mycotoxin determination. CEN has produced a document that provides criteria for mycotoxin methods (3). This document gives information about method performance, which can be expected from experienced analytical laboratories. In addition, application of analytical quality assurance procedures is recommended, including the use of certified reference materials, especially when a high degree of comparison and accuracy is required.

Socioeconomic Factors

Regulations preferably should be brought into harmony with those in force in other countries with which trade agreements exist. In fact, this approach has been applied in both the European Union (EU) and the *Mercado Comun del Sur* (MERCOSUR), i.e. the South American Free Trade Zone, where harmonization of regulations is under way. Strict regulative actions may lead importing countries to ban or limit the import of commodities, such as certain food grains and animal feedstuffs, which can make it difficult for exporting countries to find or maintain markets for their products. For example, the stringent regulation for aflatoxin B_1 in animal feedstuffs in the EU (4) led European animal feed manufacturers to switch from peanut meal to other protein sources to include in feeds; this had an impact on the export of groundnut meal in some developing countries (5). The distortion of the market caused by regulations in importing countries may lead to export of less contaminated foods and feeds, leaving those inferior foods and feeds, which do not meet the standards for export, for home consumption.

The regulatory philosophy should not jeopardize the availability of some basic commodities at reasonable prices. Especially in developing countries, where food supplies are already limited, drastic legal measures may lead to lack of food and to excessive prices. It must be remembered that people who live in these countries cannot exercise the option of starving to death today in order to live a better life tomorrow.

In the U.S. the FDA initially put into effect a regulatory policy limiting total aflatoxins in all susceptible commodities to the lowest level analytically practical (initially 30 μg/kg, later revised to 20 μg/kg) (6-8). Subsequently, studies showed that aflatoxin levels substantially higher than 20 μg/kg were not detrimental to the health of certain food-producing animals and would not result in significant aflatoxin residues in human foods derived from these animals under specific operations.

PROTOCOL FOR ESTABLISHING REGULATIONS

Risk management is primarily a product of characterized risk and the socioeconomic considerations outlined above. For the mycotoxins currently considered most significant (aflatoxins, ochratoxin A, patulin, and some trichothecenes including deoxynivalenol, fumonisins, and zearalenone), the Joint Expert Committee on Food Additives (JECFA) of the Food and Agriculture Organization (FAO) and the World Health Organization (WHO) has recently evaluated the hazard for each toxin. The latest comprehensive JECFA document on mycotoxin risk assessment, dealing with several of these mycotoxins, appeared recently (9). The qualitative indication that the contaminant could adversely affect health (hazard identification), the qualitative and quantitative nature of the adverse effects (hazard characterization), and the occurrence of mycotoxins in various commodities and food intake were embodied into each data set. Reliable data on occurrence also allow for an estimation of the effects the regulations would have on the availability of the foods and feeds concerned.

Toxicological evaluation carried out by JECFA normally results in the estimation of a provisional tolerable weekly intake (PTWI) or a provisional tolerable daily intake (PTDI). Use of the term "provisional" expresses the tentative nature of the evaluation, in view of the paucity of reliable data on the consequences of human exposure at levels approaching those with which JECFA is concerned. In principle, the evaluation is based on determination of a no-observed-effect level (NOEL) in toxicological studies and application of an uncertainty factor. The uncertainty factor means that the lowest NOEL in animal studies is divided by 100 (a product of 10 for extrapolation from animals to humans and 10 for variation among individuals) to arrive at a tolerable intake level. In cases in which data are inadequate, a higher safety factor is used.

In the 1990s, activities within the European Commission resulted in reports on the exposure assessment of aflatoxins and ochratoxin A (10, 11). The aflatoxins and ochratoxin A are mycotoxins of serious concern to the EU, where specific regulations for these toxins in various foodstuffs now exist (12). Similar activities on human exposure assessment in the EU have been carried out for patulin (13), and they are under way for the fumonisins, zearalenone, and some of the trichothecenes, including deoxynivalenol.

CURRENT MYCOTOXIN REGULATIONS

Several overviews of mycotoxin limits and regulations have been published (14–16). FAO Food and Nutrition Paper 64 published in 1997 (15) was based on a comprehensive survey of more than 100 countries that were requested, through the Agricultural Representatives of the Netherlands Embassies, to respond to a number of questions about mycotoxin regulations in their respective countries. The questions concerned the existence of mycotoxin regulations; the types of mycotoxins and products for which regulations are in force, together with maximum permissible levels; the authorities responsible for control of mycotoxins; the use of official and published methods of sampling and analysis; and the disposal of consignments that contain inadmissible amounts of mycotoxins. Responses were received from 41 countries. This was supplemented with information from other sources, so data became available from a total of 90 countries.

The respondents provided mycotoxin regulations as they existed by October 1, 1996. Most new information came from South American countries. Of the 90 countries, 77 were known to have some mycotoxin regulations, and 13 countries had no specific regulations. No data were available for about 50 countries, most of them in Africa (Figure 21.1). Most of the existing mycotoxin regulations concern aflatoxins in foods. In fact, all countries with mycotoxin regulations have regulatory levels at least for aflatoxin B_1 or the sum of the aflatoxins B_1, B_2, G_1, and G_2 in foods and/or animal feedstuffs. Less frequently, specific regulations also exist for aflatoxin M_1 in milk and milk products. Table 21.1 summarizes worldwide aflatoxin regulations in existence in 1987 and 1996. Many countries regulate the aflatoxins with limits for the sum of aflatoxins B_1, B_2, G_1, and G_2, sometimes in combination with a specific limit for aflatoxin B_1. It is debatable whether a regulatory level for the sum of the aflatoxins, which requires more analytical work than for aflatoxin B_1 alone, contributes significantly to better protection of public health than a regulatory level for aflatoxin

B$_1$ alone. Aflatoxin B$_1$ is the most important of the aflatoxins, considered from both toxicology and occurrence viewpoints. It is unlikely that commodities will contain aflatoxins B$_2$, G$_1$, and G$_2$ and not aflatoxin B$_1$, whereas the concentration of the sum of aflatoxins B$_2$, G$_1$, and G$_2$ is generally less than the concentration of aflatoxin B$_1$ alone. Typical occurrence ratios for aflatoxins B$_1$ and B$_2$ (mainly produced by *Aspergillus flavus*) average about 4:1. Typical occurrence ratios for aflatoxin B$_1$ and aflatoxins B$_2$, G$_1$, and G$_2$ (the G toxins are produced mainly by *Aspergillus parasiticus*) average about 1:0.8, although variations do occur for both ratios. Monitoring agencies in those countries that apply a regulatory level for the sum of the aflatoxins should inspect their analytical data to see how frequently the availability of data on the sum of the aflatoxins (above that on aflatoxin B$_1$) has been indispensable to adequately protect the consumer. FDA's current action levels for aflatoxins in the United States and the European Commission's current maximum admissible levels for aflatoxins in the EU are shown in Tables 21.2 (17–19) and 21.3 (20), respectively.

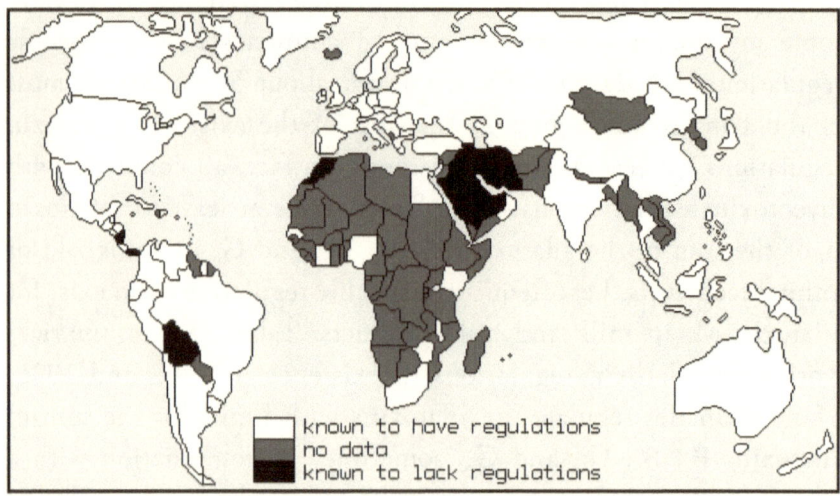

Figure 21.1. Countries known to have regulations (white), not known to have regulations (black) and where status of mycotoxin regulations is not known (gray)

Table 21.1. Medians and ranges in 1987 and 1996 of maximum tolerated levels (ng/g) for some (groups of) aflatoxins and numbers of countries that have regulations for them

Aflatoxin	1987			1996		
	Median	Range	No. of countries	Median	Range	No. of countries
B1 in foodstuffs	4	0-50	29	4	0-30	33
Total aflatoxin in foodstuffs	7	0-50	30	8	0-50	48
B1 in foodstuffs for children	0.2	0-5	4	0.3	0-5	5
M1 in milk	0.05	0-1	13	0.05	0-1	17
B1 in feedstuffs	30	5-1000	16	20	5-1000	19
Total aflatoxin in feedstuffs	50	10-1000	8	50	0-1000	21

For several other mycotoxins, specific regulations exist as well (i.e., patulin, ochratoxin A, deoxynivalenol, diacetoxyscirpenol, zearalenone, T-2 toxin, chetomin, stachybotryotoxin, phomopsin, and fumonisins B_1 and B_2). Although the fumonisins are a major target now in surveillance programs, regulations are still scarce. It is expected though that several countries will soon regulate the fumonisins, especially as JECFA now has evaluated their toxicity.

Some free trade zones (EU, MERCUSOR) are in the process of harmonizing the limits and regulations for mycotoxins in their respective member states. In the EU, this harmonization has included a tightening of the limits, in particular for the aflatoxins (20), which will lead to further debates between the EU, the United States, and the developing countries in organizations such as the World Trade Organization and Codex Alimentarius. It is not likely that worldwide harmonized limits for mycotoxins will soon be within reach.

At the time of this writing, an update of FAO Food and Nutrition Paper 64 was in preparation. The updated document, scheduled to appear in print in 2003, describes the situation on worldwide regulations of mycotoxins as they exist in 2002 (12).

Table 21.3. Maximum admissible levels for aflatoxins established in the European Union by the European Commission.

Commodity	Maximum Content (µg/kg)[a]
Milk	0.05
Peanuts, nuts, and dried fruits (direct human consumption)	2/4
Peanuts (to undergo physical processing before human consumption)	8/15
Cereals (for direct human consumption or to undergo physical processing before human consumption)	2/4
Cereals (to undergo physical processing before human consumption)	4
Spices	5/10
Feed materials (except as listed below):	50
Peanuts, copra, palm kernela, cottonseed, babassu, corn, and products derived	20
Complete feedingstuffs for cattle, sheep and goats (except dairy cattle, calves, and lambs)	50
Complete feedingstuffs for calves and lambs	10
Complete feedingstuffs for dairy cattle	5
Complete feedingstuffs for pigs and poultry (except young animals)	20
Other complete feedingstuffs	10
Complementary feedingstuffs for cattle, sheep, and goats (except dairy animals, calves, and lambs)	50
Complementary feedingstuffs for pigs and poultry (except young animals)	30
Other complementary feedingstuffs	5

[a]*Maximum contents for human foods are expressed both for aflatoxin B_1/total aflatoxins B_1, B_2, G_1, and G_2, except for milk, where the maximum content concerns aflatoxin M_1. Maximum contents for animal feedingstuffs are expressed for aflatoxin B_1.*

CONCLUSIONS AND RECOMMENDATIONS

Regulations for mycotoxins have been established in many countries to protect consumers from the harmful effects mycotoxins may cause. Most of the regulations focus on aflatoxins. Various factors play a role in the decision-making process of setting limits for mycotoxins. These include scientific considerations such as level of human exposure, toxicological data, and analytical methodology. Economic and political factors such as commercial interests and sufficiency of food supply have their impact as well. By 1997, at least 77 countries had specific regulations for mycotoxins, 13 countries were known to have no specific regulations, and no data were available for about 50 countries, many of them in Africa. Over the years, a large diversity in tolerance

Levels for mycotoxins has remained. Some free trade zones (EU, MERCOSUR) are in the process of harmonizing the limits and regulations for mycotoxins in their respective member states; it is not expected that worldwide-harmonized limits for mycotoxins will soon be within reach, as this harmonization is a slow process because of the different views and interests of those involved. Whereas harmonized tolerance levels would be beneficial from the point of view of trade, one might argue that this would not necessarily be the case from the point of view of (equal) protection of consumers around the world. The risk arising from mycotoxins is a product of the toxicity of these toxins and the level of exposure to them. The hazard of mycotoxins to individuals is probably more or less the same all over the world, although other factors may play a role as well (e.g., hepatitis B virus infection in relation to the hazards of aflatoxins). Exposure is not the same because of differences in levels of contamination and food consumption patterns in various parts of the world.

The number of countries with specific regulations for mycotoxins has been increasing over the years, and this will also be visible in the update of FAO Food and Nutrition Paper 64 (12). This reflects the general concern governments have about the potential effects of mycotoxins on the health of humans and animals. Differences are seen, however, in the legal limits that countries have laid down in their regulations. Why these limits were chosen is often not apparent.

It is recommended that:

- National governments and regional communities encourage and fund activities that contribute to reliable exposure assessment of mycotoxins in their regions.

- Efforts involving hazard assessment be coordinated and funded at the international level. Chronic toxicity studies carried out under good laboratory practice conditions are time-consuming, very expensive, and not necessarily bound to certain regions. These studies should be done in internationally recognized centers of excellence, and their results should be evaluated by international groups of experts (e.g., JECFA).

- Enacted regulations and those under development should be the result of sound cooperation between interested parties, drawn from the scientific sector, from the ranks of consumers, from industry, and from official circles. Only then can realistic mycotoxin legislation be achieved.

REFERENCES

(1)   Smith, J.W., Lewis, C.W., Anderson, J.G., & Solomons, G.L. (1994) *Mycotoxins in Human and Animal Health*. Technical Report, European Commission, Directorate XII: Science, Research & Development, Agro-Industrial Research Division, EUR 16048 EN, Brussels, Belgium

(2)   AOAC (2000) *Official Methods of Analysis*,17th Ed., AOAC INTERNATIONAL, Gaithersburg, MD, chapter 10

(3)   CEN (1999) *Food Analysis-Biotoxins-Criteria of Analytical Methods of Mycotoxins*, CEN Report CR13505, Comitée Européen de Normalisation, Brussels, Belgium

(4)   CEC (1991) Commission of European Communities, *Off. J. Eur. Commun.* **L60**, 16

(5)   Bhat, R.V. (1999) *Mycotoxin Contamination of Foods and Feeds*. Working Document. Third Joint FAO/WHO/UNEP International Conference on Mycotoxins, 3–6 March, 1999. MYC-CONF/99/4a. Tunis, Tunisia

(6) Park, D.L., & Stoloff, L. (1989) *Regul. Toxicol. Pharmacol.* **9**, 109

(7) FDA (1970) *Mycotoxins in Foods,* FDA Compliance Program R-2283, March 10 1970, Food and Drug Administration, Rockville, MD

(8) FDA (1974) Food and Drug Administration, Fed. Reg. 39, 42748

(9) WHO (2001) *Safety Evaluation of Certain Mycotoxins in Food,* Prepared by 56th Meeting of the Joint FAO/WHO Expert Committee on Food Additives. WHO Food Additives Series 47, World Health Organization, Geneva, Switzerland

(10) EC (1997) Report on Tasks for Scientific Cooperation. Report of Experts Participating in Task 3.2.1. *Risk Assessment of Aflatoxins.* Report EUR 17526 EN, Directorate-General for Industry, Office for Official Publications of the European Communities, European Commission, Luxembourg

(11) Miraglia, M., & Brera, C. (2001) Report of the EU, Scientific Co-operation on Questions Relating to Food. Task 3.2.7: *Assessment of Dietary Intake of Ochratoxin A by the Population of EU Member States (Final Draft).* European Union, Rome, Italy

(12) Van Egmond, H.P., & Jonker, M.A. (2003) *Worldwide Regulations for Mycotoxins in Food and Feed, the Situation in 2002.* FAO Food and Nutrition Paper, in press

(13) Majerus, P., & Knapp, K. (2002) Report of the EU, Scientific Cooperation on Questions Relating to Food. Task 3.2.8: *Assessment of Dietary Intake of Patulin by the Population of EU Member States (Final Draft).* European Union, Berlin, Germany

(14) Krogh, P. (1977) *Pure Appl. Chem.* **49**, 1719

(15)   FAO (1997) *Worldwide Regulations for Mycotoxins 1995. A Compendium.* FAO Food and Nutrition Paper 64, Food and Agriculture Organization, Rome, Italy

(16)   Van Egmond, H.P. (1999) *Worldwide Regulations for Mycotoxins.* Working Document. Third Joint FAO/WHO/UNEP International Conference on Mycotoxins. MYC-CONF/99/8a, 3–6 March 1999. Tunis, Tunisia

(17)   FDA (1996) *Whole Milk, Low-Fat Milk, Skim Milk–Aflatoxin M_1.* 683.100. Compliance Policy Guide CPG 7106.10, Sec 527.400, Food and Drug Administration, Rockville, MD

(18)   FDA (1996) *Foods—Adulteration with Aflatoxins.* Compliance Policy Guide CPG 7120.56, Sec 555,400, Food and Drug Administration, Rockville, MD

(19)   FDA (1996) *Action Levels for Aflatoxins in Animal Feeds.* Compliance Policy Guide CPG 7126.33, 384, Sec 683.100, Food and Drug Administration, Rockville, MD

(20)   EC (2001) Commission Regulation (EC) No. 466/2001, *Off. J. Eur. Commun.* **L. 77,**

22. PERSPECTIVES ON MYCOTOXIN REGULATIONS IN THE UNITED STATES

Henry Njapau, Garnett E. Wood and Terry C. Troxell

Center for Food Safety and Applied Nutrition
Food and Drug Administration
5100 Paint Branch Parkway
College Park, Maryland 20740

The importance of mycotoxins as human food and animal feed contaminants has been a subject of intense research and scientific discourse over the last 40 years. Mycotoxins are a food safety concern because they adversely affect the health and productivity of humans and animals. In the United States, the Food and Drug Administration (FDA) is the federal government's primary consumer protection agency for food. The FDA is responsible for enforcing a number of statutes passed by the U.S. Congress; one of the most important is the Federal Food, Drug, and Cosmetic Act. This statute and its amendments serve as the legal basis for regulating poisonous and deleterious substances in foods. Mycotoxins are regulated under Section 402 (a)(1) of the Food, Drug, and Cosmetic Act, which considers a food to be adulterated if it contains any poisonous or deleterious substance that may render it injurious to health.

The FDA has managed the risk posed by mycotoxin contamination through the risk analysis process, which is based on scientific risk assessments as well as by establishing regulatory control programs for certain mycotoxins. The risk assessment process for mycotoxins started in the mid-1960s with aflatoxins and has been modified as more information became available and other mycotoxins were discovered. Some factors that influence regulatory decisions involving mycotoxins include the type of mycotoxin, the availability of data on the incidence and levels of the mycotoxin in food commodities, the availability of analytical and sampling procedures, the availability of toxicological data, and the effect regulations would have on the availability of the food and feed.

The FDA regulatory control program for mycotoxins includes establishing regulatory limits and monitoring programs to ensure compliance with those limits; initiating appropriate enforcement action against violators; and providing guidance to the food industry on issues such as decontamination procedures, mechanisms for diversion or disposal of violative commodities, and taking the steps necessary to prevent problems or situations that might expose the public to food hazards.

ESTABLISHMENT OF REGULATORY LIMITS

Establishing regulatory limits is one way to manage the risk posed by mycotoxins. See Tables 22.1–22.4 for the FDA's current action and guidance levels. The FDA believes that preventing mycotoxin formation is the best means of managing hazards associated with mycotoxin contamination. Clearly, this is not always possible. Preharvest mycotoxin formation can be partially controlled through good agricultural practices, and postharvest fungal infection and mycotoxin formation can be significantly minimized by appropriate storage management. Once mycotoxin contamination has occurred, however, a safety management program must be established.

The FDA is a science-based regulatory agency that has put forth a comprehensive regulatory control program for mycotoxins for many years. In the establishment of various "action levels" and "guidance documents" for controlling exposure to mycotoxins, the agency used the principles of risk assessment, risk management, and risk communication. Risk assessment is a scientific process that uses

Table 22.1. FDA regulatory guidelines: action levels[a] for total aflatoxins in foods and feeds[b]

Commodity	Concentration (µg/kg)
All products, except milk, designated for humans	20
Corn for immature animals and dairy cattle	20
Corn and peanut products for breeding beef cattle, swine, and mature poultry	100
Corn and peanut products for finishing swine	200
Corn and peanut products for finishing beef cattle	300
Cottonseed meal (as a feed ingredient)	300
All other feedstuffs	20
Milk	0.5[c]

[a]Action levels are used as a guide by FDA field staff to determine when it may be necessary to take enfocement actions.
[b]FDA Compliance Policy Guides, Sections 527.400, 555.400, and 683.100 (1996).
[c]Aflatoxin M_1.

Table 22.2. FDA regulatory guidelines: action level[a] for patulin in apple juice[b]

Commodity	Concentration (µg/kg)
Apple juice, apple concentrates, and apple juice products based on the leve found or calculated to be found in single-strength apple juice or in the single-strength apple juice component of the product	50

[a]Action levels are used as a guide by FDA field staff to determine when it may be necessary to take enforcement actions.
[b]FDA Compliance Policy Guide, Chapter 5, Subchapter 510 (2001).

qualitative and preferably quantitative information to determine the risks associated with particular hazards. In this regard, the U.S. policy on regulation of mycotoxin contamination in food and feeds has been shaped by consideration of several factors, including knowledge of the incidence and levels of mycotoxin contamination of specific agricultural commodities, the inherent biological properties of mycotoxins, and the epidemiology of mycotoxin-related diseases in various parts of the world.

Table 22.3. FDA regulatory guidelines: advisory levels[a] for deoxynivalenol in wheat, derived products, and other grains[b]

Commodity	Concentration (mg/kg)
All finished wheat products that may be consumed by humans: e.g., flour, bran, and germ	1
All grains and grain by-products destined for ruminating beef and feedlot cattle older than 4 months and four chickens (these ingredients should not exceed 50% of the diet of cattle and chickens)	10
All grains and grain by-products destined for swine (these ingredients should not exceed 20% of the diet)	5
All grains and grain by-products for all other animals (these ingredients should not exceed 40% of the diet)	5

[a]Advisory and guidance levels are guidelines provided to food producers by the FDA that represent the best estimation of the negligible risk level based on available exposure and toxicological information
[b]Letter to State Agricultural Directors, State Feed Control Officials and Food, Feed and Grain Trade Organizations from R.G. Chesemore, Associate Commissioner for Regulatory Affairs, FDA, Sept. 16, 1993.

Table 22.4. FDA regulatory guidelines: guidance levels[a] for fumonisins in human foods and animal feed[b]

Commodity	Total fumonisins µg/g ($FB_1 + FB_2 + FB_3$)
Degermed dry-milled corn product	2
Whole/partially degermed dry-milled corn product	4
Dry-milled corn bran	4
Cleaned corn intended for masa production	4
Cleaned corn intended for popcorn	4
Corn and corn by-products for:	
Equids and rabbits	5 (<20% of diet)
Swine and catfish	20 (<50% of diet)
Breeding ruminants, poultry, mink, lactating dairy cattle, and laying hens	30 (<50% of diet)
Ruminants > 3 months old raised for slaughter and mink raised for pelt production	60 (<50% of diet)
Poultry raised for slaughter	100 (<50% of diet)
All other species or classes of livestock and pet animals	10 (<50% of diet)

[a]Advisory and guidance levels are guidelines provided to food producers by the FDA that represent the best estimation of the negligible risk level based on available exposure and toxicological information.
[b]FDA. Mycotoxins in Domestic Foods. Compliance Program Guidance Manual, 7307.001. Feed Contaminants Compliance Program, 7371.003.

Furthermore, in establishing action levels and guidelines, the agency was aware of the inherent assumption that the food commodities would henceforth be considered or presumed safe. Action levels were not established to be considered a tolerance or a no-observed-effect level for the various toxins; they were established as informal guidelines for use by FDA's field staff in carrying out enforcement activities.

A historical narrative illustrates the sequential thought process and perception of FDA scientists as they approached the revelation that aflatoxin, then isolated from peanuts, was injurious to health. In the early 1960s, when aflatoxin was isolated from peanut meal and shown to be an animal liver carcinogen, the FDA, the peanut industry, and the U.S. Department of Agriculture (USDA) developed practical analytical methods and created a voluntary agreement between peanut growers, sellers, and the USDA for inspection of peanuts for aflatoxin contamination and certification of raw peanut lots as either positive or negative for aflatoxin. It was not certain at that time whether aflatoxins would cause liver cancer in humans in the United States; however, the reports of severe carcinogenic effects in experimental animals and positive correlations between dietary aflatoxins and primary human liver cancer in other parts of the world were sufficient justification to regard aflatoxins as poisonous or deleterious substances and to take action to hold the human exposure to aflatoxins in the United States to the lowest level possible. By 1965, the FDA, was sufficiently confident with the available analytical methodology for aflatoxins and issued an informal action level of 30 µg/kg for total aflatoxins in peanut products. In 1969, the action level was lowered to 20 µg/kg and it was broadened to cover all foods.

In 1974 the Commissioner of the FDA considered all the information available on aflatoxins and sought to bring into balance the following four factors: the need to minimize human exposure to aflatoxins; the capabilities of sampling procedures and analytical methods to detect, measure and confirm aflatoxins; the capability of agriculture and manufacturing technology to prevent and remove contaminated peanuts; and the need for continued availability of a low-cost protein source (i.e., peanuts). An assessment of the risk associated with aflatoxins in consumer peanut products and other food commodities was later developed, thereby giving the FDA the flexibility to manage risk under the Act. The risk assessment showed that the 20 µg/kg action level adequately protected the health of consumers. Enforcing acceptable aflatoxin levels minimizes human exposure to mycotoxins and at the same time provides for an adequate food supply. The regulatory limits set for animal feed considered both the health of the animal and potential residues in foods derived from animals fed mycotoxin-contaminated feed.

ESTABLISHMENT OF MONITORING PROGRAMS

In establishing mycotoxin-monitoring programs, it was necessary to identify which mycotoxin(s) and product(s) and/or commodities should be included. This was governed by prior knowledge of the commodities susceptible to specific mycotoxin contamination in the field and in storage. Monitoring is a necessary component of any regulatory program that is focused on naturally occurring toxins. The monitoring programs yield information on incidence, levels, and geographic distribution of the mycotoxins of interest. In the United States, monitoring for common mycotoxins in grains, nuts, and oilseeds and their manufactured products is routinely carried out by the FDA as well as other federal agencies, various states, and food producers.

Monitoring ensures that foods and feeds are of an expected quality and safety. It also builds a scientific basis to determine the feasibility of avoidance. Commodities found to be highly contaminated may be diverted toward other lower risk uses. Monitoring can also be directed at specific commodities or used to alert regulatory officials of potential risk to society as a whole or to specific groups within a community. Young, old, pregnant, and immunocompromised individuals are more vulnerable to disease than normal consumers and would benefit from a diet less laden with mycotoxins. Furthermore, monitoring exercises are not restricted to domestic products, the FDA routinely inspects food imports. Mycotoxin surveillance in imported foods is necessary considering the variable or nonexistent regulations in a number of exporting countries. Shipments found in violation are detained at the ports of entry and the particular exporter must demonstrate that future shipments are not contaminated.

ALTERNATIVE MANAGEMENT PROGRAMS

Risk management options have advantages and disadvantages. Prevention is the option with minimal disadvantages. Once food is contaminated, the options become limited. For example, in the case of the mycotoxin patulin, which is produced by fungi in rotten apples, measures such as sorting to remove rotten or damaged apples from the production line are often sufficient and more practical than discarding a product that exceeds the U.S. patulin action level of 50 µg/L in apple juice. The preventative action of sorting shifts the patulin in juice distribution curves to lower incidence and levels, thus minimizing the risk of loss of revenue to juice producers that may result from compliance with the action level. This process also reduces overall exposure in addition to eliminating a product that exceeds an action level.

In other instances and for other mycotoxins, application of pesticides, including fungicides, may be a viable option. There is also the possibility of producing genetically modified (GM) crops. Available data have shown that such crops have significantly less aflatoxin contamination. In addition, some GM food crops may have a higher yield than traditional varieties, a very attractive incentive. Thus, the use of GM food crops may potentially minimize mycotoxin contamination as well as alleviate hunger in some developing countries. There is much controversy about the use of GM crops around the world, however, and it will be some time before they are widely accepted.

Chemical and physical decontamination has also been investigated and shows promising results for animal feeds. One chemical procedure that has proven effective is the high-pressure/high-temperature ammoniation technique. The procedure permanently degrades 99% of aflatoxin in cottonseed and cottonseed meal. Currently, the FDA is investigating a simpler and more versatile atmospheric pressure/ambient temperature variant of the ammoniation processes to decontaminate cottonseed.

INTERNATIONAL OUTREACH AND COOPERATION

Food safety is a global challenge that requires concerted international cooperation in the areas of surveillance, improving national food safety control systems, strengthening food control laboratories to monitor contaminants, and risk assessment. The United States has made substantial resource investments in this effort. This workshop is one of several global activities currently supported by the United States. Several agreements, such as the World Trade Organization Agreements on Sanitary and Phytosanitary Measures (SPS) and Technical Barriers to Trade (TBT) that promote greater harmonization and transparency in the establishment of food regulations that facilitate trade and protect the consumer, require the support of all nations.

The Codex Alimentarius is a vehicle for establishing standards and its standards, guidelines, and other recommendations are considered fair and as benchmarks for food safety. The Joint FAO/WHO Expert Committee on Food Additives (JECFA) provides evaluations based on sound scientific and risk assessment principles and provides risk assessment advice to Codex on contaminants. The United States has contributed significantly to the efforts of JECFA and the Codex Committee on Food Additives and Contaminants. For instance, the FDA and USDA helped lay the scientific foundation of the risk assessment and sampling plans that were the basis of the 15-µg/kg maximum level for aflatoxins in peanuts. Also, the United States has been very active in the last several years in establishing a code of practice to reduce mycotoxins in grains. It is anticipated that, on the international scale, maximum levels for mycotoxins in foods will be set to protect public health, facilitate trade, and facilitate the development of an integrated federal/state/international food safety system. Codex standards facilitate trade for all countries; for less developed nations, it is vital that their economies benefit from such activities.

RISK COMMUNICATION

The public is generally poorly informed about certain facts pertaining to the risk posed by consuming mycotoxin-contaminated foods. The scarcity of information becomes a risk in itself if it extends beyond the general public to farmers and food producers. For prevention measures to succeed, the farmer and the processor should be the most informed about factors that lead to the occurrence of mycotoxins in the commodities they produce and process. In the United States, the FDA and other responsible agencies have worked closely with the food industry to instill knowledge not only about prevention but also about the economic impact of mycotoxin contamination.

The situation may not be similar in all countries. What is crucial, however, is that a realistic way should be established to communicate the risk posed by mycotoxins. At the prevention level, farmers and producers need to be informed about the danger of mycotoxins and how to prevent their occurrence in an understandable manner and form. The transfer of important information dealing with good agricultural and manufacturing practices can play an important role in reducing the exposure of mycotoxins hazards to the consumer. Codes of practice are available—for instance, those put out by the Food and Agriculture Organization of the United Nations. Such efforts need to be upgraded to ensure that information is presented and delivered in a clear and easily understandable form to all persons involved in food production. It may even be worth the effort to translate such codes of practice into local or native languages. Rural farmers must be able to use such information easily. For such an effort to succeed, however, everyone must be involved—i.e., United Nations agencies, the United States and the European Union, and governments of all affected countries. It is expected that the participants gathered for this workshop will translate and implement what they learn here to their respective countries. Currently, most governments and consumers in developing countries do not have enough information about the risks posed by mycotoxins, which results in only marginal efforts being aimed at minimizing mycotoxin contamination.

CONCLUDING CONSIDERATIONS

Globally the right to adequate and safe food should be the goal. With the world population likely to reach 10 billion by 2050, the task is indeed daunting. Mycotoxin contamination of food commodities, if left unabated, impinges on the right to safe food. Prevention through preharvest procedures and post-harvest management strategies is the best method for controlling mycotoxin contamination. Furthermore, control programs, including the establishment of regulatory limits, monitoring programs, and decontamination procedures, can minimize these hazards to the consumer. However, such efforts should be based on sound risk assessment. The depth of any risk assessment is in some way dictated by the level of knowledge. It can be used to identify gaps in the knowledge database, direct regulation, and direct management options.

In today's global marketplace, dealing with food safety problems in the context of only one country is not enough. In this regard, the FDA's food safety conceptual framework is beyond safeguarding the U.S. consumer alone; it reaches out to the world at large, realizing that when all countries have equivalent and effective food safety systems in place, food, regardless of the source, will be safe. Furthermore, continued internationalization of the food business will drive the development of agreements among nations to provide assurances that their food exports are safe. International food standards established through Codex will be crucial and the development of internationally recognized regulations and control measures for mycotoxins that protect public health and promote fair trade at the international level must be vigorously pursued.

In summary, we must keep thinking globally, engage in long-range planning, and anticipate problems so that they do not become crises.

23. IMPACT OF MYCOTOXINS ON TRADE

Julie G. Adams

International Programs & Technical Affairs
Almond Board of California, and Scientific Committee
International Tree Nut Council
1150 9th Street, Suite 1500
Modesto, California 95354

This chapter provides a different perspective on the issues that surround mycotoxins, particularly from the viewpoint of their impact on trade. Food safety is a key issue that dominates global markets and the benchmarks have varied over the years. Twenty years ago, no one thought twice about the safety of food—it was assumed to be safe. However, recent food-borne illness incidences, from bovine spongiform encephalopathy (BSE) to dioxin, in a number of countries have raised consumer awareness of the possible risks associated with consuming certain foods. With the rapid electronic dissemination of information and borderless news, consumers are constantly bombarded with reports that are not always accurate. These reports create perceptions that are long lasting and difficult to correct.

Food commodities from countries and regions throughout the world are under increasing scrutiny—particularly those countries where climate and production practices are more conducive to mycotoxin development. The result is that trade can be hindered by shipment detentions and additional testing. How can we best minimize mycotoxin contamination of agricultural commodities without unduly restricting international trade for economically challenged countries? The impact of today's food safety regulations on global commerce and the need for internationally recognized mycotoxin sampling and testing procedures are discussed.

FOOD SAFETY REGULATIONS AND GLOBAL COMMERCE

From a regulatory standpoint, mycotoxin limits are established to protect consumers and build confidence in the food supply system. A number of factors, including the key issues of safety and consumer perceptions, affect the approach regulators take. The desire for rigorous testing and verification has encouraged enforcement agencies to write and enforce regulations across the supply chain (from farm to table). Rapidly improving analytical methods detect mycotoxins at levels that were previously undetectable, thereby increasing the perception that there is a likelihood of harm arising from consuming such commodities. Unfortunately, agricultural practices have not advanced at the same pace to meet the new standards. Food today presents no more of a risk—we simply know more about what is in it.

The aflatoxin-in-peanuts model is frequently used as a precedent to establish mycotoxin sampling plans, analytical procedures, and regulatory limits in the international market. Such plans have often been applied to other commodities without considering the inherent differences between commodities and/or mycotoxins. Not all commodities and mycotoxins possess similar characteristics that make them amenable to the protocols established for peanuts and aflatoxin. The occurrence and distribution of contamination can vary dramatically from one commodity to another.

There is not a universally applicable approach to sampling, testing procedures, and establishing aflatoxin limits. Even if the aflatoxin limits were uniform across countries, the differences in sampling and testing would have a major impact on the percentage of food lots rejected. For example, consider an evaluation that was done for several different protocols for aflatoxin sampling and testing in peanuts. The Dutch Code was an approach that stipulated multiple testing using a total 30-kg sample which was then subdivided, and each individual subsample having to meet the specified limit. It was estimated that this approach would result in 48.5% of the lots tested being rejected. Another approach was developed in consultation with a Food and Agriculture Organization of the United Nations (FAO) and specified a single test with a 20-kg sample. This approach was estimated to result in the rejection of 23% of the lots tested. Similarly, a United Kingdom procedure, also a single test but with a smaller sample size than the FAO protocol, was estimated to result in a 22% rejection rate.

Each of these approaches is designed to minimize the likelihood of aflatoxin-contaminated products entering consumer markets, but the risk of rejections is not necessarily shared equally. To balance the risk to the exporter and the importer when adopting an aflatoxin limit and sampling protocol for peanuts destined for further processing, the Codex Alimentarius recognized that the level of mycotoxins could be reduced at each stage of processing. The Codex hence adopted the FAO protocol as an international standard for peanuts destined for further processing. The Codex protocol differs from the approach taken by the European Union (EU), which adopted the more rigorous multiple-testing approach for consumer ready-to-eat nut products. The EU approach favors a lower risk to consumers despite the higher risk of rejection of good lots. The implications of these risk management approaches, particularly for producing countries, are significant from a commercial point of view.

Therefore, it is crucial that countries be aware of the basis for establishing mycotoxin limits and analytical methodologies. Certainly, science should be the foundation of the decision-making process in ensuring the public's health. However, it must be realized that science is not the only factor that influences food policies and regulations. Perceptions, rather than true consumer health implications, can persuade regulators to set lower limits and more rigorous procedures for testing, which can have dramatic consequences on global trading for economically challenged countries. Political and consumer priorities for food safety influence the way limits are viewed and enforced. Of particular concern is that regulators set stricter limits when there are insufficient data upon which to base a decision. Unfortunately, science is rarely absolute—at some level there is always a risk.

From a commercial standpoint, producers are dealing with bulk agricultural commodities that are subject to variable production capabilities and seasonal differences. Each customer and importing authority has different commodity priorities, sampling procedures, rejection criteria, and acceptability of origin certification. Few importing countries consider procedures at origin as equivalent and/or reliable, so there is always a risk that exported products will not be accepted upon receipt. Furthermore, customers in importing countries apply the same consumer-ready product standards to raw bulk commodities often in the name of "due diligence", without considering the further processing that will take place. With such rigorous standards becoming commonplace, exporters face an increasing risk of having their products rejected.

The magnitude of the potential impact of mycotoxins on international trade is evident when you consider the crops that are susceptible. Based on FAO 2000 data, it was estimated that world trade in food commodities susceptible to mycotoxin contamination was over $100 billion. This may be a slight overstatement, but it indicates the scope of the trade and the significance of the monetary implications of mycotoxin contamination. Furthermore, the affected commodities are grown and exported from every major region of the world. Given the sporadic nature of mycotoxin contamination as well as the fact that agricultural and climatic conditions vary dramatically, contamination can affect international trade at any time and in any place regardless of the production area or production practices.

FUTURE PROSPECTS

The list of commodities that are susceptible to mycotoxin contamination includes many of the key food and export crops produced in economically challenged countries. In addition to the previously adopted limit for peanuts destined for further processing, Codex has recently adopted a limit for aflatoxin M_1 in milk. It is important to note that the milk standard has implications for feedstuffs for dairy cattle. In June 2003, the Codex Commission will consider recommendations for patulin limits. Other Codex discussions are beginning to focus on ochratoxin A and fumonisins. It is clear that the focus on mycotoxins and their effect on trade will continue into the foreseeable future.

As we move forward, we must be mindful of the perception that the food industry is technologically capable of meeting the rigorous mycotoxin standards being set or proposed. Electronic sorting machines, mechanized packaging equipment, and automatic scanning in food processing industries can help identify and reduce the number of mycotoxin-contaminated raw commodities. However, this technology is not available worldwide. In the real world, facilities exist where almonds are hand-shelled and walnuts are cracked with a mallet. Exporters in economically challenged countries have to meet the same rigorous standards that are being established for technically advanced countries in global trade.

To move forward with the management of mycotoxins in the food supply, it is particularly important to examine what tools are available to producers in economically challenged countries. Given the implications of mycotoxin control strategies, it is critical that a clear and consistent approach be adopted. One key tool is the analytical laboratory. All laboratories should be using recognized, certified methodologies. Laboratory staff must be appropriately trained; sampling/testing procedures need to provide the accuracy, repeatability, and reproducibility that is required. Currently, there is minimal recognition of origin certificates for plant-based foods. If origin certification is to increase, mycotoxin test results obtained in exporting countries must be consistent with the levels being found by an importing authority. Many countries have mutual recognition agreements for meat-processing facilities but not for plant foods. Another issue is the ability to track food production and handling throughout the system so that it is possible to identify where and how a problem occurs and take the appropriate corrective action. Commercial operators and importing countries should be confident that the tracking system identifies the point/source of contamination.

CONCLUDING REMARKS

It is clear that, in the process of evaluating testing programs and sampling procedures, one must consider the impact of these procedures on trade and look beyond aflatoxin limits. For example, with rigorous interpretation of current regulations, a consignment of nuts destined for Europe that contains 2.0 µg of aflatoxin per kg would be accepted, whereas a consignment that tested 2.1 µg/kg would be rejected. Is the rejected consignment truly any less safe or could this be due to analytical variability? The implications on trade could be staggering. The cost of the rejection includes demurrage while waiting for the test results, replacing the consignment, returning the shipment to origin, and either reconditioning or diverting that consignment to nonfood uses if it is not possible to bring it within standards.

A fair chance to compete—as companies, as countries, and as people— is possible only if we establish credible, verifiable mycotoxin control programs. In all these decisions, a strong science base is absolutely critical in establishing regulations. The only common ground we have is the science. However, we all must recognize that science will not be the only factor in decision-making. It is important to try to minimize any subjectivity and build a consensus in setting the standards. The international standards that will be established in the future are useful only if they are based on science and truly reflect the global consumption patterns and have evaluated the true risk to consumers (not just perceived risks). Setting a low standard because a mycotoxin can be detected, rather than because it truly presents a health risk, will only result in more trade distortions. A balance must be struck between the risk to producers, the needs of consumers, our environment, economic stability, and technical capability. This is why it is so important to become an active participant in Codex. Understanding its role and decision-making process is essential for successful export of agricultural produce. Codex is founded on the principle that international standards be based on two criteria: ensuring the health of consumers and facilitating international trade. Using the guidance provided through expert consultations to assess risk is the foundation upon which standards and risk management strategies should be based. In this regard, it is absolutely critical that economically challenged countries be involved in the process at the regional coordinating committee level and in the international bodies where standards are being drafted.

Finally, it is important to realize that there are several stakeholders involved in food delivery. There must be cooperation among the researchers who conduct studies and do the analyses; the government, which establishes the limits; and industry, which ultimately delivers food to consumers. It is only by working together in partnership that we can ensure consumers receive wholesome healthy food products. In the global world in which we live, food knows no borders. It is important to ensure that, as that food moves throughout the world, consumers have confidence in its safety and that countries, regardless of their economic challenges, are fully able to participate in global trade.

ACKNOWLEDGEMENT

The author is grateful to Henry Njapau and Carole J. Shore (USFDA), and Thomas B. Whitaker (USDA/ARS/NCSU) for their valuable input and suggestions.

24. CREATING AN ACTIVE LEARNING ENVIRONMENT

Carole J. Shore

Center for Food Safety and Applied Nutrition
Food and Drug Administration
5100 Paint Branch Parkway
College Park, Maryland 20740

Adults learn best when they are actively involved in the learning process. To create an active learning environment, instructors need to depend less on the traditional lecture method for imparting information and more on discussions, laboratory experiences, case studies, games, and other hands-on experiences. The following eight steps (1) used in developing an active learning environment are featured in this paper.

Step 1: Define the instructional problem

Step 2: Determine the trainee's background

Step 3: Define performance objectives

Step 4: Specify subject content to support each learning objective

Step 5: Develop learning activities

Step 6: Select media

Step 7: Identify required supportive services

Step 8: Evaluation

DEFINE THE INSTRUCTIONAL PROBLEM

The first step is to define the instructional problem. According to the Food and Agricultural Organization (2), at least 25% of the world's food crops are contaminated with mycotoxins. The long-term objective of the International Workshop on Mycotoxins is to reduce human and animal exposure to mycotoxins through increased awareness of health risks associated with mycotoxin contamination; accessibility to training and detection methods; knowledge of conditions leading to mycotoxin formation; regulation and monitoring programs; and compliance with international trade standards. In defining the instructional problem, you will analyze the learning needs of the trainees, list the desired competencies trainees should have, and list the principles and concepts that will be used to meet their learning needs.

DETERMINE THE TRAINEE'S BACKGROUND

Next the instructor must address the knowledge, skills, and attitudes of each trainee. Ask the following questions: Do the trainees have the necessary educational/experience background preparation for your instruction? Are they already proficient in what will be taught? This information is critical for planning appropriate performance objectives and training experiences. If your audience consists of a combination of beginners and advanced trainees, plan to use the advanced trainees as mentors while the newer people practice a skill.

Also, ask these questions: Why do they need to learn this information? How will they use this information in their jobs? Adults learn best when the material you provide helps them solve a current problem. You need to know how the new skills and information will be used in their workplace.

One way to assess entry knowledge or skills is to give a pretest well before instruction. It is preferable to use the test as a screening device for selecting trainees. After you identify the trainee's knowledge of the subject matter, you can plan appropriate learning experiences. Many instructors, however, administer a pretest at the beginning of their training. This allows limited time for the instructor to make adjustments to their instruction and learning activities. If a post-test is administered, the results of the pre-test can be compared with the post-test to indicate the gain(s) in knowledge/skills. Another way to assess entry skills is to require prerequisite training before your instruction.

DEFINE PERFORMANCE OBJECTIVES

Once you have identified their training needs, you need to develop the objectives.

A performance objective (also called instructional objective) is a precise description of the observable skills or behaviors expected of a trainee after instruction. The behavior must be observable so that the teacher and trainee both know when the objective has been reached. Performance objectives help you decide the following:

1. What, exactly, is to be learned?

2. Who is to learn it?

3. How well is it to be learned?

4. How can you best teach it? And

5. How do you assess it?

Therefore, stating your performance objectives can improve:

1. Communication between instructor and trainee;

2. Evaluation of trainee's skills, knowledge, and attitudes before, during, and after instruction;

3. Design of instruction by determining the specific content to provide, sequence of information to be presented, and materials and methods to be used; and

4. Accountability of teaching: how many trainees achieved the objectives?

Three components must be stated in the objective: terminal behavior, test conditions, and standards of performance. Behavior refers to observable actions and movements. The objective should be a statement of a trainee's behavior and not a statement of what the instructor is going to do. Use an action verb (Table 24.1) to denote the behavior. Where the behavior is covert, such as a thought process or a feeling, state the objective in such a way that it describes what the trainee will do to infer the covert behavior. The following objectives do not indicate observable behavior:

1. The student understands thin-layer chromatography and high performance liquid chromatography.

2. To comprehend ....

How can you determine the students' understanding? One way is to state what they will be doing to show they understand. For example, *the student explains in writing the difference(s) between thin-layer chromatography and high performance liquid chromatography*. In this example, the observable behavior is stated with the action verb "*explains*." In Table 24.1, you will find this verb under the comprehension heading. Here the emphasis is on the meaning, intent, and relationships between concepts.

The thre domains or categories of objectives are:

- Cognitive: concerned with behaviors demonstrating intellectual skills, knowledge, and information, such as naming, listing, solving, identifying.

- Affective: concerned with behaviors that relate to feelings, emotions, and attitudes, such as enjoying, sensing, trusting.

- Perceptual motor: concerned with behaviors that require coordination of the brain and skeletal muscles such as typing, driving, athletic games, manipulating.

Table 24.1. Classification of verbs related to a cognitive taxonomy

Knowledge: emphasis on recall, specific or universal

VERBS

choose	answer question	complete a word, phrase, or statement
define	label	record
identify	list	confer (to gain information)
review	locate	copy
survey	match	read
select	indicate	

Comprehension: emphasis on grasp of meaning, intent, relationships in oral, written, graphic, nonverbal communication

VERBS

classify	interpret	convert
describe	measure	compare the importance of
estimate	recognize	put in order
expand	suggest	compute
explain	summarize	review to explain
express	trace	

Application: emphasis on applying appropriate principles or generalizations

VERBS

arrange	discuss	perform activity
apply	implement	plan activity
calculate	coordinate	prepare
construct	make	present
draw	solve	use information, tools
demonstrate	schedule	collect information
differentiate	keep records	administer test
		compile data

Analysis: emphasis on breakdown into constituent parts and the way they are organized

VERBS

analyze	review to analyze	make inferences
debate	form generalizations	organize
determine	deduce	interpret relationships
differentiate	draw conclusions	

Synthesis: emphasis on putting together elements or parts to form a whole

VERBS

design	combine and organize	write (original writing)
develop	plan program	coordinate (program design)
produce		

Evaluation: emphasis on values, making qualitative or quantitative judgment with criteria from internal or external sources and with standards

VERBS

evaluate	make a decision	compare and contrast

Adapted from Bloom, B.S., et al. (1969) *Taxonomy of Educational Objectives Handbook I: Cognitive Domain*, David McKay Company, New York, NY.

Most of the objectives you will use are in the cognitive domain. Bloom's *Taxonomy of Educational Objectives* (3) uses six levels—knowledge, comprehension, application, analysis, synthesis, and evaluation—to rank the verbs in terms of complexity. The lowest level is the knowledge level, the next lowest level is comprehension, and the highest level of cognitive skills is evaluation. To "apply" a concept, the trainee must first have "knowledge" and "comprehension" of the concept. Thus, the lower levels of Bloom's *Taxonomy* are the building blocks for the higher levels. For example, if you want a trainee to "analyze" the sampling plans for mycotoxins in grains, the trainee must first have "knowledge" and "comprehension" of how to sample lots of grain and must "apply" these principles in conducting inspections.

For the test condition(s), you should describe the situation in which the student will be required to demonstrate the behavior and what tools, books, raw materials, or other essentials will be provided. Sometimes the test condition(s) indicates how much time is allocated to perform the behavior. The test condition may also specify where the test will take place such as on the job, in a simulator, or in a class or laboratory. The test condition for the objective is italicized below.

Given a microscope and 10 prepared slides, the student will identify all the different species of fungi contained on the slides.

The performance standard for the objective describes the minimum level of performance that will be accepted as evidence the learner has achieved the objective. Examples of performance standards and an objective are listed below:

1.   ... must correctly answer 9 of the 10 questions

2.   ... must complete the engine tune-up so that RPM & DWELL are within 2% of factory specifications.

3.   ... must be accurate to ±1 cm.

4. Given a list of six nutrients and a list of six functions, the student will match the proper function to each nutrient.

"*The student will know the health risks associated with mycotoxin contamination.*" Is this a well-written performance objective? No! The objective lacks specificity. It does not indicate how trainees are to demonstrate what they know and to what extent they know the required information. The level of performance expected should be stated as a part of the objective unless perfection is implicit or unless the level of performance expected of the trainee is made clear to the student in other ways. Other conditions, such as time limitations and the use of additional materials or references, should be clearly specified in the objective (4). Following is a rewrite of the previous objective that informs trainees what they are expected to do to demonstrate that learning has occurred: "*The student will list and describe in writing at least four health risks in humans associated with mycotoxin contamination as discussed in the textbook.*"

Writing objectives is a developmental activity that requires changes, additions, and refinements. It is recommended to start with loosely worded objectives, move ahead in the planning sequence, and return to spell out the objectives in detail as each one becomes more evident. Remember to state what is to be learned (the behavior) and how achievement will be measured (test conditions and standards of performance). Your objectives will help you determine the adequacy of your training. One word of caution: Just because a trainee can tell the teacher how he/she would do something does not mean that he/she can do it, that he/she will do it in a different situation, or that he/she can do it with acceptable accuracy and speed. Telling how to do something and doing it are two different behaviors.

SPECIFY SUBJECT CONTENT AND DEVELOP LEARNING ACTIVITIES

Some trainers begin the process of developing a training program by specifying the subject content. This is not a recommended practice! Your performance objectives, along with the definition of the instructional problem and trainee's background, will determine the subject content. In many training sessions, the textbook is often the primary instructional resource for content. Again, this practice is not recommended. The textbook should only be considered as one resource and other resources should be incorporated into the training session.

In planning and developing learning activities, answer the following questions: What will the student do? What will the teacher do? What activities will be done to achieve the objective?

Lectures are appropriate for passive learning and are frequently used in the education of children. Adults want to be actively involved in their training. Consider using a modified version of the traditional lecture called participative lecture. This type of lecture is suited for groups of no more than 35 people and provides two-way communication between the trainers and the participants. Using a participative lecture format allows trainers to quickly identify the trainee's understanding of the subject. This technique can also be used to lead the group to make desired conclusions. One disadvantage of participative lectures is that the discussion may become irrelevant and time-consuming unless the trainer controls the questions and discussions. Some hints in preparing lectures are:

- Be prepared. Anticipate questions trainees may have.

- Schedule time for questions.

- Keep lectures short (20 minute rule).

- Vary the pace.

- Mix lectures with other learning activities.

- Use stories to apply information.

Adult learners bring a variety of experiences to the training session. They want concrete, realistic, doable practices. To keep their attention, offer practical options and workable approaches to common problems. This means you should minimize theory and maximize practice of skills. Plan for discussions and let them challenge your ideas. The trainer and other participants can learn from shared experiences. Serve as a resource instead of a traditional "teacher." Most adults want the "big picture," not a step-by-step approach to learning. They want to apply what they learn now, not in the future. Every now and then, let the trainees know how they will benefit from what you are teaching them. Offer simulations or laboratory experiences to demonstrate their understanding of the material presented.

Telling is not teaching. Often presenting information in a lecture or showing trainees how to perform a procedure is not the most effective way to deliver information. This is especially true if you are trying to engage trainees in problem-solving activities or specific psychomotor tasks. In these cases, use simulation or hands-on activities (5). To address the different adult learning styles, it is important to present a variety of activities and use various training aids to ensure that all participants succeed in learning. Those who are verbal/linguistic benefit from oral communication, puzzles, word games, and writing activities. Visual learners may be confused by the constant flow of words in sentences and paragraphs, but they can understand a concept more easily if it is graphically illustrated (Figure 24.1). For verbal learners, a visual helps to deepen their understanding and broadens their abilities to communicate. There are also bodily/kinesthetic learners. These people learn best when their bodies are actively involved in the learning process. Concepts are easier for them to learn if they are part of a song or rhyme. Offer hands-on experiences and kinesthetic learners will perform well (6).

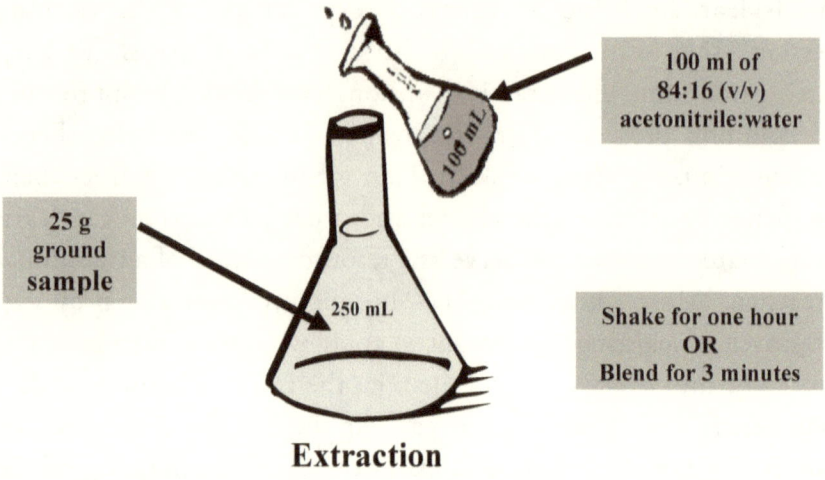

Extraction

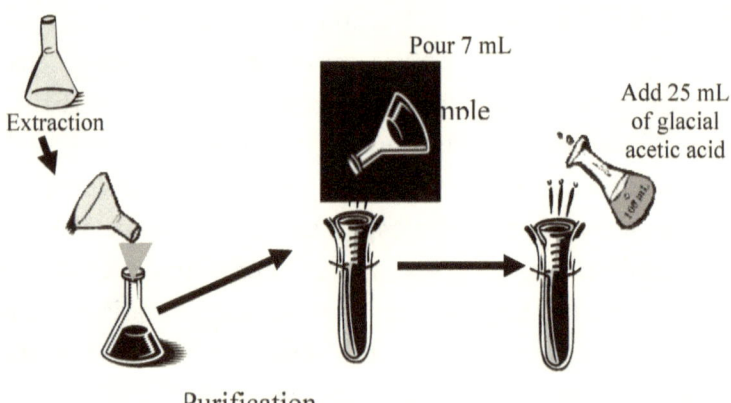

Purification

Figure 24.1. Method for Aflatoxin, DON, and Zearalenone (3-toxin test)

The first 30 minutes of the first day of training is very important. It is when participants begin to form opinions about the trainer and the training course. Trainers need to establish a climate that makes participants comfortable, relaxed, and eager to learn. One way to accomplish this is to conduct a warm-up exercise. Many trainers start each day with a warm-up exercise to stimulate thinking and participation and/or to get participants to relax. One warm-up exercise (also called an ice breaker) is to introduce yourself and your background. You may ask them how they best receive information. Are they visual or do they prefer written or oral information? This information may be useful in the delivery of other planned activities. Another warm-up exercise is called "mistaken identities." Prepare a name tag for each participant ahead of time. As the participants enter the room, randomly give each person a name tag. Then ask participants to circulate around the training room looking for the person the name tag really belongs to. Let them give the name tag to the appropriate person. Continue until most participants have the correct name tag. It works well in large groups and helps participants make new acquaintances. Allow 10–15 minutes for this exercise. When warm-up exercises are used, trainers must:

- Give directions so participants clearly understand what they are to do,

- Identify time limits and adhere to them,

- Monitor the activity by providing assistance or encouragement if needed, and

- Process the outcomes of the exercise before moving to the next activity.

Your action verb in the objective suggests the types of activities (see activity recommendations in Table 24.2) you should offer. Learning activities should be interactive and appropriate for adult learners. Whenever possible, the training content and activities should be sequenced according to the way they are performed on the job. Build from simple to complex learning tasks. Sequence modules/activities in logical steps. Provide time for self-evaluation and correction.

Table 24.2. Appropriate activities for various cognitive domain levels

Knowledge: ability to recall, list, bring to mind the appropriate materials

Possible teaching techniques:	Reference reading
	Working in groups to prepare lists
	Lectures
	Pictures
	Films, television
	Programmed instruction

Comprehension: ability to restate or explain in one's own words; recognize examples of idea without seeing fullest meaning

Possible teaching techniques:	Demonstration
	Pictures
	Models
	Group discussions
	Films, television
	Field trips

Application: ability to use concepts, facts, and principles in new, unfamiliar, concrete, or theoretical situations

Possible teaching techniques:	Games
	Simualtion
	Role playing
	Case studies
	Laboratory
	Films, television

Analysis: ability to break down facts, processes, objectives, etc., into constituent parts; to make organization of parts clear

Possible teaching techniques:	Case situations
	Simulation
	Laboratory
	Debate
	Demonstration

Synthesis: ability to put together parts and elements into a unified, original organization or whole

Possible teaching techniques:	Write and essay
	Drawing
	Prepare case situation

Evaluation: ability to make personal, original judgment on the value of ideas, procedures, methods, etc., by using appropriate criteria

Possible teaching techniques:	Problem situation tests
	Role playing
	Observing case study

The average U.S. training material is written at a sixth grade reading level. This does not mean you should underestimate the intelligence of your target audience. What it means is that it is important to write clearly, concisely, and as simply as possible. You do not want to waste valuable time in the classroom tripping over long sentences or trying to figure out what something means. Written information handouts serve a very useful purpose. They provide a record of important concepts that you want trainees to learn. Instead of focusing on the act of recording the discussions, the trainees are free to pay attention to the meaning of what is discussed. Here are some tips on developing training materials:

- Break material down into sections and subsections.

- Label sections and subsections appropriately.

- Use bullets to present lists or key points.

- Avoid long paragraphs.

- Write simple sentences.

- Delete unnecessary words.

- Use a simple, easy-to-read font.

- Summarize each module and provide a preview of the next module to ensure continuity.

- Edit.

SELECT MEDIA AND IDENTIFY SUPPORT SERVICES (PERSONNEL, TIME, SCHEDULES, FACILITIES, BUDGET, EQUIPMENT, MATERIALS)

There are no clear-cut guidelines for selecting training aids (media). The media should reinforce your instruction and not be used as the sole means to deliver information. Well-designed training aids should stimulate interest, increase attention, promote understanding, and provide experiences that may not be available in other ways.

Recent studies have shown that, when verbal and visual messages are combined, audiences retain nearly 50% of the information presented. The retention rate of verbal-only presentations is about 10% (7). The primary advantage of spoken narration is in its effectiveness in personal and interpersonal communication. Written narration, unlike audio, allows readers to set their own pace. Visuals (charts, graphs, diagrams, pictures) can make audio narration easier to follow and easier to recall later. Be sure you use only visual aids that are necessary to achieve your objective. Ask yourself whether the aid will help to explain, reinforce, or highlight your point. Visual aids need to be simple.

A flipchart is an excellent tool for brainstorming and reinforcing key concepts in small to medium-sized (fewer than 25 people) group activities. This teaching aid encourages interaction and involvement of the audience. To ensure that your handwriting is legible, prepare most of the text on the chart before class. Lightly pencil in the "preferred" responses. When the response is given, use a marker to trace over the penciled text.

EVALUATION

In designing successful training programs, instructors should also address the following questions: Why? What? How? How will you know? Your needs assessment conducted in the first step of the process answers the Why question. Steps 2 (determine the trainee's background) and 3 (define performance objectives) address the What question. How will you meet the objectives? Steps 4–7 provide guidance for this question. Trainers often overlook the important question: How will you know? Establish a plan for evaluating both your training program and the trainees' performance. Evaluation is the quality control for the process and outcomes of training. Before the training program starts, you should determine what information is needed to assess the effectiveness (How) of the training. Is the How meeting the What for the Why (8)?

Throughout the training session, you should be evaluating the trainees' progress in mastering the objectives. If trainees are having difficulty understanding or applying a concept, you need to individualize the instruction for them. Consider their learning style and other ways of explaining a concept.

Offer remedial sessions for those in need of additional instruction. At the end of the training session, most trainers will administer a test to judge the trainees' performance in learning the material presented. Three characteristics of a good criterion referenced test are validity, reliability, and objectivity. A test is valid when it requires a learner to perform the same behavior under the same conditions specified in the performance objectives. A valid test is a mirror image of the objective. A test is reliable when it provides a consistent measure of the trainee's ability to demonstrate achievement of the objective(s). The best way to increase reliability is to increase the number of opportunities the trainee has to demonstrate achievement. For example, ask a student to repeat the thin-layer chromatography (test) five times with different samples. If two or more competent observers independently agree that a trainee's test performance meets the criterion stated, then the test is objective.

If 80% of your trainees pass a given test on the first try, you can be reasonably sure the instructional materials and procedures used for that particular unit were effective and do not need revision. This assumes that the trainees did not possess the tested knowledge/skills before their training. On the other hand, if 80% of your trainees fail a given test on the first try, you should examine your instructional materials and teaching method(s) to determine what revisions are needed to improve your teaching effectiveness.

Participative lectures, discussions, laboratory experiences, and question-and-answer sessions provide good feedback on the level of understanding of the material. At the end of a training session, it is a good idea to get the trainees' evaluation of the training materials/media, experiences offered, length of time spent on various subjects, your delivery of information, etc.

By following these eight steps, your chances of having a successful training program are enhanced. The first three steps—defining the instructional problem, the trainee's background, and performance objectives—are critical activities and provide the bases for the remainder of the steps. Always apply the golden rule:

- Tell them what you are going to tell them.

- Tell them.

- Tell them what you have told them.

Keep the environment interactive and offer lots of time to practice skills learned. Remember, all programs can be improved. What works well for one audience may not be as affective for another audience.

REFERENCES

(1)   Charles, C.L., & Clarke-Epstein, C. (1998) *The Instant Trainer. Quick Tips on How to Teach Others What You Know*, McGraw Hill, New York, NY

(2)   Boutrif, E., & Canet, C. (1998) *Rev. Méd. Vét.* **149**, 681

(3)   Bloom, B.S. (Ed) (1972) *Taxonomy of Educational Objectives. Handbook 1: Cognitive Domain*, David McKay Co., New York, NY

(4)   Hough, J.B., & Duncan, J.K. (1970) *Teaching: Description and Analysis*, Addison-Wesley Publishing, Reading, MA

(5)   Backes, C.E. (1997) *Adult Learn.* **8** (3), 29

(6)   Brougher, J.Z. (1997) *Adult Learn.* **8** (4), 28

(7)   Department of Health and Human Services, Administration for Children and Families (2001) *Training of Child Support Enforcement Trainers. Trainer's Guide*, Department of Health and Human Services, Washington, DC

(8)   Gottman, J.M., & Clasen, R.E. (1972) *Evaluation in Education. A Practitioner's Guide*, F.E. Peacock Publishers, Itasca, IL

Index

A

Action Levels 326, 335, 337
Afflamummon danielli 80
Aflatoxicosis 21, 73, 76, 89, 90, 92, 98, 292, 297
Aflatoxin 22, 24, 26, 32, 34, 37, 40, 52, 53, 54, 55, 74, 76, 77, 78, 79, 80, 89, 90, 91, 92, 93, 98, 117, 118, 119, 120, 121, 122, 124, 125, 126, 131, 133, 134, 135, 136, 139, 142, 147, 154, 163, 187, 190, 191, 193, 199, 201, 202, 203, 212, 217, 218, 225, 226, 227, 230, 241, 244, 249, 250, 251, 252, 253, 254, 255, 257, 262, 265, 269, 270, 271, 272, 273, 274, 275, 276, 277, 287, 297, 299, 301, 302, 310, 312, 313, 314, 322, 323, 325, 326, 327, 328, 337, 338, 339, 340, 344, 345, 347, 348
  absorption 26, 53, 188, 193
  action levels 326, 335, 337
  ammoniation 26, 253, 254, 257, 297, 339, 340
  biological control 249, 255
  chemical inactivation 304
  detoxification 53, 248, 253, 256, 257, 298
  distribution 22, 38, 64, 96, 118, 119, 121, 133, 134, 148, 166, 199, 253, 275, 276, 277, 293, 304, 338, 339, 344
  liver cancer 23, 32, 52, 53, 54, 89, 91, 93, 337
  production in
    corn 13, 15, 30, 32, 43, 44, 74, 86, 94, 114, 130, 137, 138, 145, 149, 249, 253, 271, 272, 273, 275, 279, 282, 283, 285, 305, 334, 335, 336
    cottonseed 32, 34, 52, 115, 120, 133, 146, 202, 206, 250, 254, 297, 302, 328, 340
    peanuts 30, 32, 34, 114, 127, 130, 137, 149, 273, 275, 279, 305, 328
    tree nuts 32, 314
  toxicity 22, 51, 53, 57, 58, 59, 62, 63, 79, 85, 86, 88, 92, 93, 97, 253, 297, 310, 314, 327, 329, 330
Africa 53, 55, 58, 60, 62, 90, 93, 94, 235, 244, 251, 252, 253, 282, 325, 329
Alimentary toxic aleukia. (ATA) 56, 85, 98, 310
Almonds 34, 52, 91, 214, 250, 347
Alternaria 63, 64, 249
Ammoniation 26, 253, 254, 257, 297, 339, 340
Analysis 23, 79, 86, 116, 131, 133, 136, 139, 151, 152, 153, 154, 157, 163, 164, 165, 166, 171, 174, 177, 179, 182, 185, 186, 188, 189, 190, 191, 193, 194, 195, 196, 197, 199, 200, 202, 205, 206, 209, 210, 212, 213, 222, 231, 232, 243, 264, 265, 272, 274, 275, 276, 277, 278, 284, 285, 293, 299, 300, 301, 302, 303, 308, 309, 315, 322, 325, 333, 356
  derivatization 188, 194, 235, 237, 240, 243
  fluorimetry 195

patulin 32, 34, 55, 60, 61, 191,
213, 253, 297, 313, 323, 324,
327, 335, 339, 347
zearalenone 32, 34, 51, 56, 59, 60,
163, 191, 195, 202, 213, 222,
225, 226, 227, 230, 232, 234,
235, 239, 240, 241, 242, 243,
244, 251, 255, 274, 277, 324,
327

Ascites 86, 89

Aspergillus 32, 33, 34, 35, 36, 37, 39,
40, 41, 43, 52, 54, 55, 56, 60,
61, 86, 94, 96, 249, 251, 269,
294, 326

A. carbonarius 32
A. flavus 32, 34, 37, 40, 52, 54, 257
A. nomius 33
A. ochraceus 33
A. parasiticus 32, 33, 35, 37, 52, 86

Automated multiple development
(AMD) 212

Awareness 44, 267

B

Bacillus thuringiensis (Bt) 282, 287
Balkan endemic nephropathy 55, 310
Bangalore 92
Banswada 86, 89, 90
Barley 32, 34, 39, 51, 52, 56, 58, 61,
64, 95, 115, 120, 202, 233
Bidirectional TLC 194, 212
Bile 62, 74
Bin 133
Black pepper 97
Bloom's Taxonomy 356
Brazil 26, 94, 114, 139, 140, 150, 214,
282
Breast milk 74, 75
Bulgaria 55, 97

C

Calcium montmorillonite 79
Canara 86, 90
Cancer 23, 32, 52, 53, 54, 58, 60, 89,

91, 93, 94, 282, 310, 313, 337
aflatoxins 14, 20, 30, 32, 52, 53, 54,
74, 76, 82, 89, 91, 92, 100, 200,
205, 235, 238, 312, 331, 332
environmental factors 33, 36, 172,
256, 282, 298
mycotoxins 21, 31, 49, 50, 51, 64,
75, 77, 78, 185, 199, 202, 204,
264, 269, 291, 307, 309, 343

Carcinogen 52, 54, 55, 56, 57, 93,
312, 337

Carcinogenicity 60, 77, 89, 91, 93, 97,
311, 312

Cardiac beri-beri 61

Cashew nuts 34

Charm Sciences 205

Cheese 50, 254

mycotoxins in 2, 28, 36, 37, 38, 40, 51,
64, 73, 75, 79, 80, 85, 185, 190,
191, 199, 233, 234, 235, 242,
250, 253, 266, 269, 270, 273,
274, 275, 276, 277, 287, 292,
294, 299, 307, 308, 309, 310,
311, 314, 315, 316, 324, 327,
329, 338, 340, 341, 347, 356

Chemical Derivative Formation 219

Chemical inactivation 304

Chetomin 327

Chickens 57, 92, 336

Chilanga 138

Childhood cirrhosis 310

Chilies 91

Chittoor 86, 92

Citrinin 61, 213, 310

Claviceps 61

Claviceps fusiformis 88

Claviceps purpurea 32, 34, 61, 62, 85

Clavine alkaloid 87

Clay 79, 131, 255, 256, 297

Cleanup 156, 188, 189, 190, 191, 193,
194, 195, 204, 209, 215, 224,
227, 228

Codex 3, 117, 127, 140, 150, 151,
170, 179, 186, 273, 300, 308,
309, 310, 311, 313, 314, 315,

Thin layer chromatography 194, 232
Tibet 64
TLC 185, 186, 187, 193, 194, 195,
   209, 210, 211, 212, 213, 214,
   215, 216, 217, 218, 219, 221,
   222, 226, 227, 228, 229, 230,
   231
Tolerable daily intake (TDI) 58, 311,
   324
Toxicant 31
Toxicity 22, 51, 53, 57, 58, 59, 62, 63,
   79, 85, 86, 88, 92, 93, 97, 253,
   297, 310, 314, 327, 329, 330
   acute 32, 58, 59, 61, 66, 75, 85, 89,
      90, 93, 94, 95, 96, 98, 297, 310,
      314
   chronic 32, 53, 62, 64, 66, 74, 75,
      97, 310, 316
Toxicological determination 21, 22
Toxin 226
Trade 16, 28, 104, 151, 233, 235, 300,
   308, 323, 327, 336, 340
Trans-placental transfer 91
Transkei 58, 93, 94
Tremorgens 55
Trichoderma 58
Trichothecenes A 229
Trichothecenes B 14, 228, 229
Trier 133, 146
Turkey X 21, 52, 310
Turmeric 91, 97

U

United Kingdom 2, 91, 274, 345
United States Department of Agricul-
   ture 128
United States of America iv
Urine 57, 62, 74
Uruguay 281, 285, 302, 308
USA iii, iv, 5, 6, 7, 8, 9, 94, 104, 131,
   150, 204, 205, 209, 233, 271,
   279, 281, 291
USDA. *See* United States Department
   of Agriculture

V

Variability 117, 118, 119, 120, 121,
   123, 126, 127, 136, 174, 213,
   348
Vicam LP 6, 234, 235, 236, 243, 244
Village 136, 139, 142, 145
Vomitoxin. *See* Deoxynivalenol
Vulvovaginitis 59

W

Walnuts 52, 250, 347
Warangal 86, 92
Water activity 35
Wheat 3, 32, 34, 39, 51, 52, 56, 58,
   59, 86, 87, 88, 95, 96, 115, 120,
   163, 202, 252, 253, 274, 283,
   296, 336
World Health Organization (WHO) 2,
   52, 66, 68, 109, 166, 289, 307,
   316, 317, 318, 319, 324, 331
World Trade Organization (WTO) 28,
   151, 300, 308
Wortmannin 60

Y

Yellow rice 61, 310
Yugoslavia 55, 97

Z

Zambia 38, 137, 138
Zearalenol 60, 234, 243
Zearalenone 32, 34, 51, 56, 59, 60,
   163, 191, 195, 202, 213, 222,
   225, 226, 227, 230, 232, 234,
   235, 239, 240, 241, 242, 243,
   244, 251, 255, 274, 277, 324,
   327
Zea mays 282